Space and Time

Philosophische Analyse / Philosophical Analysis

Herausgegeben von / Edited by
Herbert Hochberg, Rafael Hüntelmann,
Christian Kanzian, Richard Schantz, Erwin Tegtmeier

Volume / Band 54

Space and Time

A Priori and A Posteriori Studies

Edited by
Vincenzo Fano, Francesco Orilia and Giovanni Macchia

DE GRUYTER

ISBN 978-3-11-055474-8
e-ISBN 978-3-11-034892-7
ISSN 2198-2066

Library of Congress Cataloging-in-Publication Data
A CIP catalog record for this book has been applied for at the Library of Congress.

Bibliographic information published by the Deutsche Nationalbibliothek
The Deutsche Nationalbibliothek lists this publication in the Deutsche Nationalbibliografie; detailed bibliographic data are available in the Internet at http://dnb.dnb.de.

© 2017 Walter de Gruyter GmbH, Berlin/Boston
This volume is text- and page-identical with the hardback published in 2014.
Printing: CPI buch bücher.de GmbH, Birkach

♾ Printed on acid-free paper
Printed in Germany

www.degruyter.com

Contents

Editors' Introduction VINCENZO FANO, FRANCESCO ORILIA and GIOVANNI MACCHIA	1
On Russell's Metaphysics of *Time* GREGORY LANDINI	7
Weak Discernibility and the Identity of Spacetime Points DENNIS DIEKS	43
A Structural and Foundational Analysis of Euclid's Plane Geometry: The Case Study of Continuity PIERLUIGI GRAZIANI	63
Can the Mathematical Structure of Space Be Known A Priori? A Tale of Two Postulates EDWIN MARES	107
Gunkology and Pointilism: Two Mutually Supervening Models of the Region-Based and the Point-Based Theory of the Infinite Two-Dimensional Continuum MILOŠ ARSENIJEVIĆ and MILOŠ ADŽIĆ	137
This Moment and the Next Moment FRANCESCO ORILIA	171
T × W Epistemic Modality ANDREA IACONA	195
Towards a Theory of Multidimensional Time Travel DOMENICO MANCUSO	209
Gödelian Time Travel and Weyl's Principle VINCENZO FANO and GIOVANNI MACCHIA	237
About the Authors	273

Editors' Introduction

Vincenzo Fano
Università di Urbino
Dipartimento di Scienze di Base e Fondamenti
vincenzo.fano@uniurb.it

Francesco Orilia
Università di Macerata
Dipartimento di Studi Umanistici—Lingue, Mediazione, Storia, Lettere, Filosofia
orilia@unimc.it

Giovanni Macchia
University of Western Ontario
Department of Philosophy, Rotman Institute of Philosophy
gmacchi@uwo.ca

Most of the papers in this collection have been discussed in a series of seminars organised in Italy in the framework of a two-year research project funded by the Italian Ministry of University and Scientific Research (MIUR) in 2008 and involving the philosophical departments of L'Aquila, Macerata, Sassari and Urbino. The main purpose of the project was to investigate the interplay of a priori and a posteriori points of view in metaphysics and to favor their interaction.

The present collection focuses in particular on the ontology of space and time and, in the spirit of the project, is centered on the idea that the difficult problems typically encountered in this area must be tackled from a multifarious perspective, paying attention to both a priori and a posteriori considerations.

In the most recent debate about the nature of space and time one might perhaps detect a tendency to separate too rigidly between a priori and a posteriori approaches to the relevant issues. For one could find that prominent scholars such as Oaklander, Le Poidevin, Nerlich, and others mainly contribute to the former kind of approach, whereas others, equally prominent, e.g., Earman, Norton and Dieks, mainly contribute to the latter kind. With the papers collected in this volume we aim at contributing to the goal of bridging the gap between these two perspectives, thereby showing the fruitfulness of combining them as much as possible. In the following we shall introduce the reader to these papers.

Deeply influenced by Einstein's relativity theory, Bertrand Russell has been most influential in combining a posteriori considerations from physics and a priori logical analysis. He may be viewed as the noble father of the currently fashionable eternalist or four-dimentionalist views of time, wherein the pastness, presentness and futurity of events is subjective pretty much like their being here or far away. Yet, a theory of this sort, with its ontology devoid of objective distinctions between past, present and future events, seems less in tune with Russell's ontologically austere "robust sense of reality" than presentism, arguably the most interesting rival of eternalist or four-dimensionalist views. Landini's contribution to this volume aims to show that this is not the case, or at least that it is not the case from Russell's own perspective.

Traditionally, one can view space and time as either independent substances (*substantivalism*) or as somehow emerging from objects and events (*relationism*). Once space and time are viewed on a par, as is the case in eternalist/four-dimentionalist accounts, the substantivalist-relationist dispute arises for space-time as a whole. Dieks' paper focuses on *structuralism* understood as a middle course between substantivalist and relationist accounts of space-time: space-time can exist independently of matter, and thus is a substance, but its properties can be fully analyzed in terms of relations. Dieks investigates what can be said from such a perspective about the identity of space-time points. In particular, analogies and disanalogies between space-time points and quantum particles are discussed, with emphasis on the status of Leibniz's Principle of the identity of indiscernibles, and the role of the notion of "weak discernibility."

Next, there is a group of four papers, due to Graziani, Mares, Arsenijević and Adžić, and Orilia, respectively, which deal with continuity, as a feature of space or time.

Graziani examines the once widely accepted idea that the continuity postulate is totally unexpressed in the *Elements* and that Euclid was totally silent on this topic. Prima facie, this classical interpretation seems to be totally correct, but an Italian scholar, Attilio Frajese, showed the limits of such an interpretation, by arguing that "although Euclid has not expressly enunciated the postulates concerning the intersections of circle and straight lines and of circle and circle he has fixed nevertheless the necessary conditions so that the intersections are effective." Graziani's paper aims to contribute to an understanding of how the *Elements* contain particular cases of the continuity postulate that are certainly used by Euclid, even though they are not explicitly stated. The postulate is probably connected to the centrality of diagrams in the *Elements*. If this is correct, it follows that, to axiomatize the geometry of the *Elements*, it is necessary to provide an account of the "synthetic procedures" implicit in the recourse to diagrams. By considering these analyses about continuity and nature of Euclidean constructions, Graziani presents a set of axioms and rules sufficient, by the modern standard, to provide a foundation of Euclid's Plane Geometry and, following contemporary developments of Substructural Proof Theory, he formalizes a Sequent Calculus for Euclid's Plane Geometry.

It may be difficult to say what the epistemological status of geometry was, according to Euclid. Reichenbach and the logical positivists were however rather explicit on this matter. They argued that even applied geometry is a priori, since we choose by convention the geometrical postulates that are meant to represent space. This conventionalism was attacked with some success by Quine, Glymour, and others. Yet, in recent years Michael Friedman has forcefully argued in favor of conventionalism. Mares' paper tackles this issue by contrasting the parallel postulate with the postulate that space is continuous. Mares comes to the conclusion that we should treat the parallel postulate as empirical and the postulate that space is continuous as a priori. Thus, according to Mares, neither party is quite right: different postulates of geometry may have a different epistemological status.

In Arsenijević and Adžić's contribution one can find an interesting formal result concerning two theories of the infinite two-dimensional continua, the region-based and the point-based theories. The result is of crucial importance for the currently widespread quarrel between gunkologists and pointilists, which is a continuation of the 'great struggle' between Aristotelians and Cantorians about the structure of the continuum. An entity, and so also any continuum, is gunky if any of its parts has proper parts that are parts of the entity itself. In this sense, the Aristotelian two-dimensional continuum is gunky, because it consists of the regions such that any of them has a region that is its proper part (and a part of the continuum itself), while the Cantorian two-dimensional continuum that consists of null-dimensional points is not gunky, since points have not proper parts at all. There is a widespread belief that the two conceptions represent alternatives to each other that are worth being further investigated. But the formal result shows that the issue of which of the two theories is true makes no sense unless we somehow extend the two theories and raise some additional questions concerning the structure of the physical world.

Orilia's contribution focuses on a version of presentism, instantaneous presentism, according to which the present is a durationless instant. Orilia discusses two prominent problems for this view. One regards the fact that we seem to have empirical evidence of events that take an extended interval of time. Orilia argues that the instantaneous presentist can account for such events by appealing to time-indexed properties such as 'was in such and such a place at a certain instant.' It is the other problem that has to do with continuity, since it consists of the Zenonian challenge, posed to the instantaneous presentist by William L. Craig, of accounting for the passage of time, given that time is continuous and thus there is no next instant. By employing, at least in part, ideas developed in dealing with the first problem, Orilia tries to counter Craig's arguments against instantaneous presentism so as to show that this doctrine is compatible with the thesis that time is continuous.

There is then a paper, by Iacona, that delves deep into logical matters in order to deal with the deteterminism-indeterminism controversy.

Iacona aims to provide a formal semantics offering a metaphysically neutral ground for the dispute between determinists and indeterminists. Iacona focuses on the so-called T×W modal frames, which allow for a treatment in one fell swoop of both modal and temporal discourse. In particular, they have been used to provide a semantics for a modal operator that expresses historical necessity, by means of the constraint that quantification over possible courses of events is limited to those that are alike up to a given point. Iacona proposes a T×W semantics without such a constraint so as to provide a system with an epistemically interpreted modal operator in order to arrive to the intended neutral ground.

Finally, Mancuso's and Fano and Macchia's essays are two papers that explore the exciting topic of time travel. In the framework of Einstein's general relativity time travels in a certain sense are physically possible. This raises both a priori and a posteriori philosophical issues.

Mancuso outlines a non-standard temporal model compatible with time travel: on the one hand, because it averts the causal loops arising in a linear chronology, and, on the other hand, because it incorporates certain formal requirements which are implicit in the way time travel is usually conceived of. The most important such requirement stipulates that each agent has permanent access to a "private past," which no one else can modify. As it turns out, the minimal model satisfying the requirements involves $n+1$ dimensions for n independent travellers, with appropriate analytic constraints on an agent's path in $\mathbf{R}^{n+1}$.

Fano and Macchia discuss an issue that the philosophical debate concerning time travels seems to have overlooked. This debate usually takes into account two kinds of time travels: *Wellsian* (WTTs) (from the H. G. Wells' literary milestone, *The Time Machine*) and *Gödelian* (GTTs) (after the logician K. Gödel, who found a model of a relativistic universe that allows for time travels). WTTs are often not considered as worth of serious scientific attention, even if they are properly travels *through* time. GTTs, on the contrary, result from physics itself, as solutions of the Einstein field equations, but they are not properly time travels. The particularity of these solutions, in fact, resides in the topological structures of their space-times, curved in such a way as to mould the so-called *closed time-like curves* (CTCs) that actually represent the very "spacetime journey" itself and not travels *over* time. However, in our universe, the existence of a *cosmic time* is deemed to be possible. Indeed, an important foundational principle of modern cosmology, *Weyl's Principle*, stating the regularity (non-vorticity) of the divergent worldlines of clusters of galaxies, allows to foliate space-time in a sequence of time slices. Universes respecting Weyl's Principle make it possible, in principle, to compare local time orders of CTCs with temporal orderings of neighbouring and distant worldlines, all existing in a sort of "absolute" temporal standard of reference. This allows us to recover for the mostly physical GTTs, at least partially, a more "natural" sense of journey *through* time, as in the case of WTTs.

ON RUSSELL'S METAPHYSICS OF *TIME*

GREGORY LANDINI
University of Iowa
Gregory-landini@uiowa.edu

ABSTRACT. Russell is famous for a four-dimensionalist neutral monism in which 'mind' and 'matter' are logical fictions. Embracing Einstein's relativity, Russell's B-theory of time rejects *becoming* and the Presentist thesis that what exists, is present. Russell's rejection of non-existing objects is no less famous. These seem in tension. The solution offered in this paper appeals to the constructivism of Russell's article "On Order in Time." All events with end points, out of which B-theorists compose series, are logical fictions constructed from *ongoing* events. There must be events with parts that are *future* to the given constructed end point of an event, and events whose parts are *past* with respect to it. Though *ongoing* events are primitive, the absence of a neighborhood entails that the notion *becoming* (with a leading temporal point) is unintelligible. Presentism's notion of the 'now' is also unintelligible. Eternalism, Growing Block, and Presentism do not exhaust the positions on the nature of time. There is yet Russell's metaphysics of *Time*.

1. Introduction

Bertrand Russell is famous for being a philosopher with a "robust sense of reality." His theory of definite descriptions, first discovered in "On Denoting" (1905), is a very important tool for the endeavor to get the truth-conditions for our sentences to square up with the correct philosophical ontology. Ramsey quipped that it is a paradigm for philosophy.[1] But what is the "correct" ontology? Quine writes:

> A curious thing about the ontological problem is its simplicity. It can be put in three Anglo-Saxon monosyllables: "What is there?" It can be answered, moreover, in a word—"Everything." (Quine 1948, 1)

The influence of Russell is apparent. Quine praises Russell for the theory of definite descriptions. Both share the view that 'existence' is not a property. Both hold the maxim: "To be is to be a value of the variables of quantification." But the appearance of agreement hides a deep and serious disa-

---

[1] Ramsey 1931, 263.

greement. For Russell, ontology derives from the attempt to offer special kinds of entities that ground necessities in fields such as arithmetic, geometry, physics, biology and the like. In Quine's view, ontology is simply about what there is; and what there is, is a matter for empirical science to decide. He writes:

> Within natural science there is a continuum of gradations, from statements which report observations to those which reflect basic features say of quantum theory or the theory of relativity. The view which I end up with [...] is that the statements of ontology or even of mathematics and logic form a continuation of this continuum, a continuation which is perhaps yet more remote from observation than are the central principles of quantum theory or relativity. The differences here are in my view differences only in degree and not in kind. (Quine 1951, 211)

The question of whether there are coelacanths in the seas off South Africa is an ontological question for Quine no less so than whether there are *numbers*. For Russell, the existence of coelacanths is a contingent empirical matter, not a matter of ontology. Their existence is not a metaphysical postulate to explain some form of necessity. Russell maintains that speculative metaphysical ontologies motivated by philosophical endeavors to ground allegedly non-logical necessities (arithmetic, geometric, causal, metaphysical, etc.) is little more than a muddle produced by unclear thinking. The only necessity is logical necessity. The correct ontology is the ontology of logic.

This interpretation of Russell might, at first, seem a surprise. What about mind and matter, the soul, and sense-data? Aren't Russell's theories of the existence of such entities ontological views? If we regard them as ontological then we use the word "ontological" in a way that is different from Russell's formal use. Russell's sense-data (at least from 1913 onward) are *physical* transitory particulars whose existence is not part of a philosophical theory to account for necessary relationships, but taken to be part of a contingent empirical theory of physics. Sense data are not postulated as the ground of a necessity, and thus are not part of an ontological view. The case of "matter" and "mind," is even more complicated. Many ordinary notions such as "matter" are hybrid notions involving a mixture of mathematical, logical and physical components. The entanglements of these notions produce the appearance that there are forms of necessity (metaphysical, causal, arithmetic, geometric) besides logical necessity.

There are many strange questions[2] concerning necessary relationships involving matter. Is matter necessarily infinitely divisible? Is everything colored necessarily extended and at a place in space? In Russell's program, the philosopher's task is to separate logical and mathematical components from physical components in a way that reveals that the only necessity is logical necessity. Once the components are distinguished, the existence of matter and mind is no longer an issue pertaining to ontology.

Russell's research program in philosophy was shared by Carnap. But he was influenced by Wittgenstein's radical Tractarian extension of Russell's program: all and only logical/semantic notions (*exists*, *physical object*, *number*, *class*, *truth*, *fact*, *representation*, etc.) are *shown* by logical grammar and not *said* in pseudo-predicates. Carnap attempted to reveal that *all* ontological/metaphysical notions, including the ontology of logic, are meaningless. Carnap's conventionalism in logic played a central role in his account. Russell's research program was transmogrified into a program of building the pseudo-predicates of analytic (necessary) truths into language forms. Quine argued convincingly against Carnap's conventionalism in logic. He reinstated the pseudo-predicates of analytic truths by adopting a holistic empiricism which abandons introspective awareness in favor of neural conditioning and a theory of meaning as translation. This behaviorist account of meaning cannot sustain a distinction *in kind* between analytic and synthetic. The pseudo-predicates of philosophical ontology Carnap wished to relegate to meaninglessness are thus no different in kind from the ordinary predicates of the physical sciences he so lauded.

Russell's position retains the traditional meaning of "ontology" as the study of the objects that ground necessary relationships. In contrast with Quine, ontology is certainly not to be placed in the hands of physics—the empirical scientific study of what there is. Neither, however, is it meaningless as Carnap thinks. In Russell's view, philosophers who engage in speculative metaphysics to account for non-logical necessities have failed to separate the logical and mathematical components from the physical components of concepts. Their metaphysical confusions were products of their poor logic (inability to get the truth-conditions right) and these led them to statements, which, by the lights of proper logic, are ill-formed (meaningless). But ontology is certainly not meaningless. Logic (and its mathematical branch) has a special status in Russell's philosophy. Schools of philosophy, Russell tells us, should be characterized rather by their logic

---

[2] See Coffa 1977.

than their metaphysics (Russell 1924, 323). The correct ontology is the ontology of logic.

The implications of Russell's scientific philosophy for understanding his views on the philosophy of time are worth exploring carefully. Let us recall John McTaggart's famous A, B and C series.[3] We have:

A-series: "[...] the series of positions running from the far past through the near past to the present, and then from the present to the near future and the far future." McTaggart declared that "the distinctions of past, present and future are essential to time and that, if the distinctions are never true of reality, then no reality is in time." He considered the A series to be 'temporal,' a true time series because it embodies these distinctions and embodies change.

B-series: "The series of positions which runs from earlier to later [...]" The B series is temporal in that it embodies direction of change. However, McTaggart argues that the B series on its own does not embody change.

C-series: "[...] this other series [...] is not temporal, for it involves no change, but only an order. Events have an order. They are, let us say, in the order M, N, O, P. And they are therefore not in the order M, O, N, P, or O, N, M, P, or in any other possible order. But that they have this order no more implies that there is any change than the order of the letters of the alphabet [...]" According to McTaggart, the C-series is not temporal because it is fixed forever.

In *The Principles of Mathematics* (1903), Russell clearly espouses a B-theory. Russell explains, in opposition to McTaggart, that a B-theorist can offer a proper analysis of change. He writes:

> The concept of motion is logically equivalent to that of occupying a place at a time, and also to that of change. Motion is the occupation by one entity, of a continuous series of places at a continuous series of times. Change is the difference, in respect of truth or falsehood, between a proposition concerning an entity and a time T and a proposition concerning the same entity and another time T', provided that the two propositions differ only by the fact that T occurs in the one when T' occurs in the other. (Russell 1903, 469)

---

[3] McTaggart 1908.

There are many passages in Russell's writings throughout his long life advocating some form of the B-theory. In *Human Knowledge: Its Scope and Limits*, he writes:

> For the moment, therefore, I take as raw material 'events,' which are to be imagined as each occupying a finite continuous portion of space-time. It is assumed that events can overlap, and that no event recurs. [...] Since we have agreed that events, so far as known to us, are not merely instantaneous, we shall wish to define 'instant' in such as way that every instant exists at a continuous stretch of the series of instants. That instants must form a series defined by means of the relation of earlier and later is one of the requisites that our definition must fulfill [...] We are thus compelled to search for a definition which makes an instant a structure composed of a suitable selection of events. Every event will be a member of many such structures, which will be the instants during which it exists; it is "at" every instant which is a structure of which the event is a member.
>
> [...] someone might object, that if the world were to remain without change for, say, five minutes, there would be no way of fixing a date within these five minutes [...] This, however, is not an objection to our statement, but only to the supposition that time could go on in an unchanging world. On the Newtonian theory this would be possible, but on a relational theory of time, it becomes self-contradictory. If time is to be defined in terms of events, it must be impossible for a universe to be unchanging for more than an instant. And when I say "impossible," I mean *logically* impossible. (Russell 1948, 169)

As we see, the B-theory (at least as Russell formulates it) makes time without change impossible, for change is defined in terms of overlapping events.

The B-theory of the metaphysics of time is now called the "Eternalist view" or "Block (Static) Theory." According to this theory, events are tenseless and occur in relations of temporal order. A more moderate view is called the "Growing Block." On this view, past and present events are accepted, but since there is becoming, no events future to the edge of becoming exist. Early advocates of the B-theory, including Russell and Reichenbach, coupled it with a thesis that tense is not semantically primitive and thus that all tensed statement are adequately translatable into tenseless statements referring to times. Thus, for example, the past tense of

"Italy won the World Cup"

is to be captured as

"The event of *Italy's winning the World Cup* is at some time before this utterance."

Russell explains that "science professes to eliminate 'here' and 'now.' When some event occurs on the earth's surface, we give its position in the space-time manifold by assigning latitude, longitude and date" (Russell 1948, 6). He goes on to write:

> I maintain that in addition to the objective relation of before-and-after, by which events are ordered in a public time-series, there is a subjective relation of more-or-less remote, which holds between memories that all exist at the same objective time. This private time-series generated by this relation differs not only from person to person but from moment to moment in the life of any one person. There is also a future in the private-time series, which is that of expectation. Both private and public time have, at each moment in the life of a percipient, one peculiar point, which is, at that moment, called "now." [...] It is to be observed that 'here' and 'now' depend on perception; in a purely material universe there would be no 'here' and 'now.' [...] We may define 'I' as 'the person attending to this,' 'now' as 'the time of attending to this,' and 'here' as 'the place of attending to this.' We could equally well have taken 'here-now' as fundamental; then 'this' would be defined as 'what is here-now,' and 'I' as 'what experiences this.' (Russell 1948, 91)

A. N. Prior 1968 argued conclusively that such translations cannot succeed since they rely essentially on the use of indexicals such as "now" or "this." An utterance involving an indexical is, according to the B-theory, yet another event. The order of this event in the B-series is not at all fixed by its being an utterance. The utterance, as an event, has no privileged status in the series. Every event has its own local time which is its *now* (present). Prior's arguments seem telling. As a result, some new advocates of the B-theory concede that tense is semantically ineliminable, but insist that this is relevant only to one's individual consciousness of time and has no bearing on the metaphysics of time itself. They maintain that all events are untensed entities.

In contrast, Presentists take *now* (*present*) to be metaphysically fundamental, and thus tense as a fundamental feature of events. Interestingly, Presentists divide over whether the quantifiers "all" and "some" express ontological commitment to *present* objects. Bourne argues that "now" should not be likened to "exists" and thus it is not to be given with quantification. In this respect, he departs from Russell's view that commitments to what is (in the mind-independent world) are given by means of the variables of quantification. Bourne worries that if "present" is given with quantifiers, then the difference between Presentism and the B-theory cannot even be articulated.[4] Presentism, which is supposed to say that what exists is present, would reduce to an "uninteresting truism" saying that whatever is, is. Bourne's concern, however, is misplaced. A Presentist who likens "present" to "exists" and thus makes it part of quantification, is charging that the B-theorist, just as the Meinongian, holds the untoward position that there are (presently exist) events of which it is true to say they are not (*present*). The B-theorist has a ready reply. The existence (present existence) of an event is not to be confused with *when* the event was unfolding (its situation in the temporal order). On the B-theory, the event of *Caesar's crossing the Rubicon* exists (presently exists), but Caesar is not presently engaged in the crossing.

There is thus no tension whatsoever in Russell's adoption of the B-theory and the thesis that "existence" (or even "present existence") is given with quantification.[5] There is, however, a fundamental tension in Russell's philosophy of time. The tension arises from Russell's constructivism in mathematics and the implications this has for his construction of events. Past and future events with *beginning points or end points* (as eternal untensed entities standing in relations of temporal order) seem to be constructions (relations on other events). This constructivism requires that some events be taken as fundamental. But the ones that are fundamental seem to be out of sorts with the metaphysics of Eternalism, the Growing Block, and Presentism. In this paper we shall try to address some of these issues confronting Russell's metaphysics of time.

---

[4] See Bourne 2006, 10.
[5] On this view, "All men are mortal" does not mean that all men, past, present or future, have always been, are, and ever shall be, mortal.

2. Mathematics and the metaphysicians

In "On Denoting," Russell wrote: "A logical theory may be tested by its capacity for dealing with puzzles, and it is a wholesome plan, in thinking about logic, to stock the mind with as many puzzles as possible, since these serve much the same purpose as is served by experiments in physical science."[6] Of course, it not just any "puzzles" Russell has in mind. It is those puzzles that rise to the level of significant paradoxes. Paradoxes might best be characterized as an apparently unacceptable conclusion derived by apparently acceptable reasoning from apparently acceptable premises.[7]

Russell came soon to hold that "Logic is the essence of philosophy." More exactly, he writes: "Every philosophical problem, when it is subjected to the necessary analysis and justification, is found either to be not really philosophical at all, or else to be, in the sense in which we are using the word, logical" (Russell 1914, 42). Russell's conception of a new scientific philosophy (a research program he called "logical atomism" was inspired by the great advances made by mathematicians and physical scientists. The new mathematics of the infinite eventuated in new logical analyses of notions such as *number, continuity, limit, space* and *time*. "Quantity," wrote Russell, "[…] has lost the mathematical importance which it used to possess, owing to the fact that most theorems concerning it can be generalized so as to become theorems concerning order."[8] Let me quote in length:

> The notion of continuity depends upon that of order, since continuity is merely a particular type of order. Mathematics has, in modern times, brought order into greater and greater prominence. In former days, it was supposed (and philosophers are still apt to suppose) that quantity was the fundamental notion of mathematics. But nowadays, quantity is banished altogether, except from one little corner of Geometry, while order more and more reigns supreme. The investigation of different kinds of series and their relations is now a very large part of mathematics, and it has been found that this investigation can be conducted without any reference to quantity, and, for the most part, without any references to number. All types of series are capable of formal definition, and their properties can be deduced from the principles of symbolic

---

[6] Russell 1905, 110.
[7] Sainsbury 2009, 1.
[8] Russell 1901, 59.

> logic by means of the Algebra of Relatives [the higher-order logic of relations]. The notion of a limit, which is fundamental to the greater part of higher mathematics, used to be defined by means of quantity, as a term to which the terms of some series approximate as nearly as we please. But nowadays the limit is defined quite differently, and the series which it limits may not approximate to it at all. This improvement also is due to Cantor, and it is the one which revolutionized mathematics. Only order is now relevant to limits. Thus, for instance, the smallest of the infinite integers is the limit of the finite integers, thought all finite integers are at an infinite distance from it. The study of different types of series is a general subject of which the study of ordinal numbers (mentioned above) is a special and very interesting branch. (Russell 1901, 59)

In Russell's view, Cantor's work on the transfinite put to rest centuries of speculative metaphysics surrounding the "infinite" and the notion of "continuity." Russell writes: "Continuity had been, until he [Cantor] defined it, a vague word, convenient for philosophers like Hegel, who wished to introduce metaphysical muddles into mathematics. . . . By this means a great deal of mysticism, such as that of Bergson, was rendered antiquated" (Russell 1945, 829).

Of course, in heralding Cantor, Russell does not mean to disparage the work of others such as Weierstrass, Dedekind, and Cauchy whose analyses also revolutionized mathematical concepts. Indeed, Russell frequently discussed their work. And Russell certainly does not mean to say that ordinal numbers (which are based on well-ordered series) are all that is needed in defining *limits* and *continuity*. Cantor's is not the only definition of continuity; and modern analysis prefers to work with Cauchy's definition. Russell's point, however, stands. All modern notions of *limits* and *continuity*, as well as metrical notions, topology, and the like, have their foundation in structures generated by *relations* ordering their fields. The notion of *order*, not *quantity*, is central to modern mathematics.

Russell would certainly agree that more than once in the history the discovery of paradox has been the occasion for major reconstruction at the foundation of thought. Galileo thought it a paradox that the function f such that $f(n) = 2n$ correlates natural numbers one-to-one and onto the even natural numbers. He thought it paradoxical since the even natural numbers are a proper subset of the natural numbers. Indeed, a function can be found that correlates the natural numbers one-to-one and onto the integers (positive and negative natural numbers). It is this:

$$f(n) = -\sin(\frac{(2n-1)\pi}{2}) \times \frac{\sin(\frac{n^2\pi}{2})+n}{2}.$$

Cantor found a one-one function f that assigns each rational $\frac{a}{b}$ to a unique natural number n. Here is such a function:

$$f(\frac{a}{b}) = \frac{a(a+1)}{2} + a + ab + \frac{b(b+1)}{2}.$$

This seems especially surprising since for each natural number $n > 0$ there is a rational $\frac{1}{n}$ such that $\frac{0}{1} < \frac{1}{n} \leq \frac{1}{1}$. At first these results seemed paradoxical, but they brought about a revolution in how we think about infinity.

Russell holds[9] that a rational $\frac{a}{b}$ is not a quantity or a physical magnitude. It is a relation of natural numbers m, n defined as follows:

$$m\,(\frac{a}{b})\,n =_{df} bm = an.$$

The identity of ratios, the addition of ratios, and the multiplication of ratios are defined thus:

$$\frac{a}{b} = \frac{c}{d} =_{df} m\,(\frac{a}{b})\,n \equiv_{mn} m\,(\frac{c}{d})\,n \quad {}^{10}$$
$$\frac{a}{b} \times \frac{c}{d} =_{df} m\,(\frac{ac}{bd})\,n$$
$$\frac{a}{b} + \frac{c}{d} =_{df} m\,(\frac{ad+bc}{bd})\,n$$

For example, $\frac{1}{2} = \frac{7}{14}$ because

$$m\,(\frac{1}{2})\,n \equiv_{mn} m\,(\frac{7}{14})\,n.$$

For all natural numbers m and n, $2m = 1n$ iff $14m = 7n$. With these definitions $\frac{m}{0}$ is allowed, but it is uninteresting and worth ignoring. It is easy to see, however, that the definition yields the result that for all $m > 0$, $\frac{0}{m} = \frac{0}{1}$,

---

[9] *Principia* offers a more general definition to cover cases such as $\frac{a}{b}(\frac{e/f}{g/h})\frac{c}{d}$ and even further embeddings. We have simplified matters here for convenience of exposition.
[10] We write $Am \equiv_{mn} Bn =_{df} (\forall m)(\forall n)(Am \equiv Bn)$.

and that $\frac{0}{0} \neq \frac{1}{1}$. We can readily prove:

If $b \neq 0$ then $\frac{a \times b}{b} = \frac{a}{1}$.

This is the law for cancellation.

On this construction, the natural number n is not identical with the rational $\frac{n}{1}$. Indeed, the constructions eliminate the ontology of rational numbers. No rational number satisfies the equation: $x \times 2 = 1$. Rather, this equation is to be replaced by another equation, viz.,

$$\frac{a}{b} \times \frac{2}{1} = \frac{1}{1}$$

which has solutions. For example, this has solutions when $a = 1$ and $b = 2$, and when $a = 7$ and $b = 14$, and so on. The "uniqueness" of the solution is preserved as follows:

$$\frac{r}{c} \times \frac{2}{1} = \frac{1}{1} \equiv_{r,c} \frac{r}{c} = \frac{a}{b}$$

All these solutions are "identical" by the definition of identity for ratios.

Russell's construction is eliminativistic. It is not a reductive identity. When we say water is H_2O, light is electromagnetism, or rust and fire are both cases of burning due to oxygen infusion we have reductive identities in mind. We say there are such entities, though they are not quite what we originally thought them to be. Elimination is quite different. Russell's eliminativism becomes clear when he speaks of rational numbers, real numbers, sets and the like as "logical fictions." In modern set theory, the rational number $\frac{1}{2}$ is reductively identified with a set of ordered pairs $<m, n>$ of natural numbers m, n such that $m(\frac{1}{2})n$. But in Russell's analysis, there are no sets, and so there is no object $\frac{1}{2}$. Talk of such entities is merely a convenience.

This form of elimination which preserves the structure of laws without the ontology of objects satisfying the laws takes some getting used to. It is a method widely used in mathematical constructions. Russell endeavored to use the method to solve philosophical problems. He writes:

> One very important heuristic maxim which Dr. Whitehead and I found by experience, to be applicable in mathematical logic, and have since applied in various other fields, is a form of Ockham's razor. When some set of supposed entities has neat logical properties, it turns out, in a great many instances, that the supposed entities can be replaced by purely logical structures composed of entities which have not such neat properties. […] The principle may be stated in the form: 'Whenever possible, substitute constructions out of known entities for inferences to unknown entities.' (Russell 1924, 326)

In *Principia*, Whitehead and Russell maintain that classes, propositions, signed integers, ratios, reals, complex numbers and the like are "logical fictions." That is, there are no classes, propositions, signed integers, ratios, reals, complex numbers and the like. Instead, the laws governing those entities are restructured and recovered. Russell's reconstructions are eliminativistic; they are not reductive identifications.

Some philosophical paradoxes raise very deep problems with mathematical, scientific and metaphysical conceptions of the world. Russell hoped to address them by applying the very same techniques used in mathematics. Many important paradoxes arise from importing intuitions wrought from experience with physical objects into the realm of mathematics, and *vice versa*. Here is an old paradox. Suppose we divide a line segment AB in half.

```
├────┼────┤
A    C    B
```

Now there is no such thing as cutting a point in half, for by definition, points have no length. But then if we are to cut a line in half, in which half is the mid-point? If the length of a line is determined by its points, then if the midpoint is in line segment CB and not in line AC, then AC is not as long as CB since it has one less point. Driven by our physical intuition, it seems that we haven't divided the line in half.

The paradox is solved by realizing, as Cantor did, that every line segment has the same cardinal number of points. In the figure (below) there is a one-to-one correspondence between point P on line segment CD and point P' on the *shorter* line segment AB.

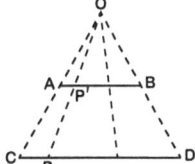

Intuitively, for any physical objects which could be associated with this figure, the sticks laying along from P to P' would have width, and each would overlap more of AB than CD. But sticks are not made up of points and sticks don't intersect at points. The mathematical conception of lines and points are *not* physical conceptions. Now the cardinal number of points on any line segment is the non-denumerable cardinal number $2^{\aleph_0}$. This is the cardinal number of the Real numbers (which is the same as the cardinal number of subsets of natural numbers). Cantor proved that $\aleph_0 < 2^{\aleph_0}$. There is no function correlating natural numbers onto subsets of natural numbers, and yet there is a function correlating each natural number with a subset of natural numbers (e.g., just correlate each n with the set $\{n\}$). The mathematical notion of comparative *length* involves the notion of a *metric* which is a relation imposing a particular structure on its field. But to get the general idea of the role of relations involved, let us take the different and simpler case of different orderings of natural numbers. Natural numbers are *well-ordered* by < (the *less-than* relation) as follows:

    0, 1, 2, 3, 4, 5, 6, …

Consider the quite different ordering,

    0, 1, 3, 4, 5, 6, 7, …; 2.

This ordering is generated by the quite different *well-ordering* relation $<^{+1}$ which is this:

    x and y are natural numbers and either $y = 2$ and $x \neq 2$ or $x \neq 2$ and $y \neq 2$ and $x < y$.[11]

---

[11] See Russell 1919, 91.

Both of these series have the same cardinal number of members, but the well-ordering relation ordering the first may be said to exact an ordering that is smaller in "length" because the first is isomorphic to a proper part of the second (and not *vice versa*). Ordinal numbers, unlike cardinals, are understood in terms of well-ordering relations. The order-type of the well-ordering relation < which orders the natural numbers is the foundation of the ordinal ω, and the order-type of the well-ordering relation $<^{+1}$ is the foundation for the ordinal number $\omega + 1$. An *order topology* can be readily defined on any well-ordered set, and thus ordinals may be used in the mathematical notion of *measurement*. The key point, however, is that *relational structure* is at the foundation of the mathematical conception of the metrical notion of *length*.

Mathematical lines and points have no width, and so we have a one-to-one correlation of points on CD with points on AB, even though the line segment AB is *shorter* than the line segment CD. The paradox of dividing a line in half is dissolved when we divorce the mathematical notions *point* and *length* (of a line segment) from intuitions based on experiences with physical pebbles and sticks.

What are "points"? Modern analytic planar geometry, following Descartes, takes them to be pairs of numbers. A "straight line," in the mathematical sense, is a mathematical function (a relation R of a special sort such that, for some numbers *m, b,* and for all numbers *x, y* either *x*R*y* if and only if $y = mx + b$ or for some number *c*, we have $mx = c$.&. $y = y$. Such points on a line (pairs of numbers) obviously have no length at all. As we saw, the mathematical notion of "length" is not determined by the number of points, but by the *ordering* of the points. That is, the ordering of the pairs of numbers by a relation R (the one line) is structurally similar to just a proper part of the ordering by the relation S (the other line). The sleight of hand that produces the paradox of dividing a line in half lies in misapplying our mathematical ideas to the physical world. The mistake is to think about lines and points the way we think about physical sticks and pebbles.

But let us return to physical sticks and pebbles. Zeno of Elea is most famous for paradoxes that arise when mathematical notions of infinity are (*mis*)applied to the world. One of Zeno's popular paradoxes is known as Achilles and the Tortoise. The argument aims to show that if, in a race, the tortoise starts some distance ahead, even Achilles (the very fastest of Greek runners) would be powerless to catch up. Achilles has to transverse a greater distance than the tortoise, so it seemed to Zeno that if he catches the tortoise at some time, he has to have consecutively transversed infinitely more measurement points than the tortoise in the same amount of time.

Of course, we can see that in the mathematical sense of "length" the distances traveled by both Achilles and the tortoise have the same infinite cardinal number of points even though Achilles's distance is greater. The difference in distance (length) is not due to a difference in the cardinal number of points, but the ordering of those points. There is a one-to-one correlation between the points of the distance Achilles runs onto those of the tortoise. The cardinal number of points on any line segment of any length is the same as that of any other. In his *Principles of Mathematics* (1903) Russell likens the situation to his paradox of Tristram Shandy, who in writing a very detailed autobiography, is ever falling further and further behind the events of his life. Nonetheless, if Tristram lives a life with as many days as there are natural numbers, there is a one-to-one correlation between his life events and entries in his autobiography.[12]

But this does not seem to fully address the Achilles paradox. After all, Achilles catches the tortoise. Tristram Shandy never completes his autobiography. Zeno argues that if the physical distance between Achilles and the tortoise is infinitely divided, Achilles has to *consecutively* transverse an infinity in a finite amount of time to reach the tortoise. If catching the tortoise is to complete an infinity in the sense of consecutively transversing each point making up the distance, then it is surely logically impossible. Now some interpret Russell as holding, to the contrary, that it is logically possible to *complete* an infinity—but Russell has in mind a quite different sense of "completion". In "The Limits of Empiricism," Russell wrote:

> Miss Ambrose says it is *logically* impossible to run through the whole expansion of π. I should have said it was *medically* impossible. [...] The opinion that the phrase 'after an infinite number of operations' is self-contradictory, seems scarcely correct. Might not a man's skill increase so fast that he performed each operation in half the time required for its predecessor? In that case, the whole infinite series would take only twice as long as the first operation.[13]

There is an equivocation in the meanings of "running through," and "operating" and "completing." It is logically impossible for there to be a *consec-*

---

[12] Thanks to Francesco Orilia for a discussion of this.
[13] See Russell 1935b.

utive operation which accumulates an infinity in a finite temporal interval. But not all operations are consecutive; the notion of a term that is positioned in a relational ordering "after an infinite number of terms" is definable.

Russell's comments certainly do not show that he maintains the logical possibility of *completing* an infinity by means of a *consecutive* accumulation. To be sure, Russell notes that the "running through the whole expansion of π is only "medically impossible," and he means that its physical impossibility is quite compatible with its logically possibility. But Russell's point about logical possibility in this passage is to argue against finitism in mathematics (which embraces only "potential infinities"). Russell accepts the "actual infinity" as a perfectly intelligible mathematical notion. The notion of a number coming "after" an infinite interval, has a perfectly determinate mathematical sense defined by appeal to the well-ordering relations. For example, there is an infinity of ratios $\frac{p}{q}$ less than the ratio $\frac{2}{1}$, and yet due to the ordering of the ratios, the rational bound $\frac{2}{1}$ occurs "after" the infinity. This is not at all mysterious. But it certainly does not involve the connotations of consecutiveness involved in Anscomb's notion of *running through* all the rationals that come before $\frac{2}{1}$.

In fact, this distinction is precisely what provides for the dissolution of Thomson's paradox of the oscillating lamp.[14] The lamp starts out *off*, but then oscillates so that thereafter it is *on* at a t if and only if it was *off* at some distinct *t-n*, and it is *off* at a given t if and only if it was *on* at some distinct *t-n*. If the oscillations are such that the first is at t, then the second in $\frac{1}{2}t$, and so on, then infinitely many oscillations have occurred in the finite interval $2t$. Thomson maintains that if such an infinity is transversed in a finite interval, then we have a contradiction, for at mark $2t$ the lamp is *on* if and only if it is *off*. But Thomson's word "transversed" connotes consecutiveness. Once that connotation is removed we see that he is incorrect. The mark 2t is not *in* the interval of oscillations; it is the least upper bound of the interval and so comes after. Thomson's lamp fails to generate a contradiction.

But even though the Thomson lamp story fails to generate a paradox for the notion of *non-consecutively* completing an infinity in a mathematical sense, it is clear enough that what Anscombe and Thomson and Zeno have in mind by "transversing" or "completing" or "running through," or

---

[14] See Thomson 1954.

'oscillating" are physical events which certainly do seem to form *consecutive* series. Zeno's paradox cannot be dispatched by appeal to the mathematical notions related to ordering. Zeno presumes that Achilles's catching the tortoise is a physical event involving a *consecutive* accumulation of an infinity of physical events. The trouble is the consecutiveness. The mathematical notion of *completion*, unlike the physical, does not connote consecutiveness.

A better approach to Zeno's paradox, therefore, is to question the presumption that consecutive accumulation of distances is involved in running (catching the tortoise). What does it mean for physical distances to be infinitely divided? Given a physical measuring technique (say repeatedly laying a stick end to end), Achilles proclaims that he has caught the tortoise at a particular time t. The tortoise, however, complains that the measuring device was inexact and that by his more refined measuring device Achilles is still a little behind him. Achilles then retorts that even on the tortoise's new refined device, he catches the tortosie at time $t+m$. The tortoise then offers a further refinement of his physical measuring device which shows that Achilles is still a little bit (smaller yet than the last one) behind. On and on this dispute goes—provided the tortoise can come up with ever more refined physical measuring techniques enabling measurements in the ratios. If physical measuring devices can be created with any refinement allowable by the denseness of the ratios, their dispute goes on without end. But no matter what measuring device the tortoise employs, Achilles is justified in saying that that device shows him catching (and then surpassing the tortoise) at some time. Zeno cannot generate a paradox this way. He cannot legitimately say: "At the infinitely small, the measuring device has Achilles not catching the tortoise." Zeno has yet to give any clear meaning of there being such a physical measuring device capturing an infinitely small physical distance.

This approach is sympathetic to Aristotle's analysis according to which physical distances, unlike mathematical lengths, are not infinitely divided. Aristotle's own concerns aside, it is clear that the mathematical notion of being divided is distinct from the notion (whatever it is) of being *physically divided*. In its place, we offered the notion that a physical distance may be understood as infinitely divided insofar as physical objects can always be found to realize measurements of any given degree of accuracy in the rationals. We found a sort of stalemate between the arguments of Achilles and those of the tortoise. Can Zeno generate a paradox from this stalemate? The tortoise has argued that for any supposed catch point t, he can find a refined physical measure which reveals that Achilles does not catch him at t. Suppose however that Achilles does run past the tortoise and choose say t^* as a surpass point. Now at t^* Achilles is ahead of the tortoise and this holds no matter how refined a measure the tortoise may choose. The tortoise first shows that Achilles catches the tortoise at $t+m_1$ and not at t. The tortoise's next more refined measure shows Achilles didn't catch the tortoise at $t+m_1$, but rather at $t+m_1+m_2$, and so forth for ever more refined measures and smaller and smaller increments m_i of time. But in any finite number of iterations of the tortoise's appeal to more refined measures, Achilles is still ahead of the tortoise at time t^*, provided that it is larger than the finite sum $t + m_1 + m_2 + , ..., + m_n$. Now Achilles might argue that ordinary intuitions of distance and space demand that since t^* is stable for every physical measure, there must be a catch time t that is *also* stable for any physical measure. The tortoise has shown that ever more carefully refined physical measurements reveal that there is no one such stable catch time t. Zeno might try to argue that since there is no such stable catch time t, physical intuitions of distance suggest there can be no stable time t^*. Zeno might well maintain that Achilles cannot surpass the tortoise if he cannot catch him at some point. Unfortunately for Zeno, these physical intuitions are not reliable.

Zeno seems perfectly content to present Achilles's consecutive accumulation of infinitely many physical distances as perfectly well modeled in terms of mathematical *addition* of infinitely many ratios representing increasingly smaller lengths of the distances. He seems perfectly content to present the accumulation of time as a consecutive mathematical addition of ratios representing the temporal durations. Zeno is content to do this because in his day the mathematical notion of summing an infinite series was unintelligible. Zeno then had the upper hand. But in the nineteenth-century, the problem of defining the mathematical notion of the arithmetic sum of an infinite series was resolved. Russell takes this new mathematics to generate the complete solution of the paradox. Consider the following:

> Zeno was concerned, as a matter of fact with three problems, each presented by motion, but each more abstract than motion, and capable of a purely arithmetic treatment. These are the problems of the infinitesimal, the infinite and continuity. To state clearly the difficulties involved, was to accomplish perhaps the hardest part of the philosopher's task. This was done by Zeno. From him to our own day, the finest intellects of each generation in turn attacked the problems, but achieved, broadly speaking, nothing. In

our own time, however, three men—Weierstrass, Dedekind and Cantor—have not merely advanced the three problems but have completely solved them. The solutions, for those acquainted with the mathematics, are so clear as to leave no longer the slightest doubt or difficulty. This achievement is probably the greatest of which our age has to boast; and I know of no age (except the golden age of Greece) which has a more convincing proof to offer of the transcendent genius of its great men. Of the three problems, that of the infinitesimal was solved by Weierstrass; the solution of the other two was begun by Dedekind, and definitively accomplished by Cantor. (Russell 1901, 64)

Suppose we give the tortoise a 9 mile head start, and have him run at 1 *mph* and have Achilles run at 10 *mph*. Then when Achilles runs 9 miles, the tortoise is $9 + \frac{9}{10}$ miles (9.9 miles) from the starting line. When the Achilles is at $9 + \frac{9}{10}$ miles (9.9 miles) from the start, the tortoise is at $9 + \frac{9}{10} + \frac{9}{100}$ miles (9.99) from the start. Again when Achilles is at $9 + \frac{9}{10} + \frac{9}{100}$ miles (9.99) the tortoise is at $9 + \frac{9}{10} + \frac{9}{100} + \frac{9}{1000}$ miles (9.999), and so on it goes if the ground can be measured in the rational numbers. The purely mathematical reply to Zeno is that motion is not properly understood as *consecutive* addition. Indeed, it concedes to Zeno that no consecutive addition will enable the arithmetic sum of infinitely many finite increments. To see this point, consider summing consecutively as follows. Begin with $\frac{9}{10}$ and then add to that sum $\frac{9}{100}$ and then add to that sum $\frac{9}{1,000}$ and then $\frac{9}{10,000}$ and so on. The astute philosopher will point out that these consecutive additions "towards infinity" do not equal 1. They cannot reach infinity (whatever that means). Fair enough. But it is a naïve philosopher who thinks that the ratio $.\overline{9}$ is not equal to the ratio $\frac{1}{1}$. The definition is this:

$$.\overline{9} =_{df} \lim_{n \to \infty} \Sigma(\frac{9}{10^{n+1}}) .$$

Weierstrass showed how to define the *arithmetic sum* of an infinite series without any appeal to consecutive additions or infinitesimal quantities. Consider the following series

$$\frac{9}{10}, \frac{9}{100}, \frac{0}{1,000}, \frac{9}{10,000}, \frac{9}{100,000}, \ldots$$

The function that determines the series (its general term) is $\frac{9}{10^{n+1}}$ since the assignment of *n* consecutively to 0, 1, 2, 3, etc., yields the series. How can one define the notion of the *arithmetic sum* of this series? Weierstrass's idea is to define the *limit* of the series as follows:

$$\lim_{n \to \infty} \Sigma(\frac{9}{10^{n+1}}) =df$$

the ratio $\frac{L}{p}$ such that for all ratios $\frac{e}{q} > \frac{0}{1}$ there is a ratio $\frac{N}{r} > \frac{0}{1}$ which is such that for all *n*, $\frac{Lq-pe}{pq} < \Sigma(\frac{9}{10^{n+1}}) < \frac{Lq+pe}{pq}$. The arithmetic sum of our series is $\frac{1}{1}$. For every choice of $\frac{e}{q} > \frac{0}{1}$ there is a band between $\frac{Lq+pe}{pq}$ and $\frac{Lq-pe}{pq}$ such that we can find an $\frac{N}{r} > \frac{0}{1}$ after which for all arguments *n*, all values of the function stay in the band. With his logical analysis of the notion of a *limit,* Weierstrass liberated the calculus from the notions of "completing an infinity" or a dynamic notion of "getting infinitely close to" and thereby reaching a quantity. In Russell's view, philosophers have paid too little attention to his breakthrough.

Let us then heed Russell's advice and take seriously Weierstrass's breakthrough. Weierstrass shows us how to define the *mathematical* notion of "summing" an infinite consecutive series *without* appealing to consecutively completing infinitely many partial additions. We may accept that well enough. But we may very well conclude that this reveals that physical motions, and temporal durations, are not properly modeled by the mathematical notion of summing an infinite series! Physical motion certainly seems to be a process of consecutive accumulation of spatial parts. The *ordinary* empirically informed physical intuition is that Achilles *consecutively* accumulates finite segments of the physical distance to the tortoise during his run. Zeno's paradox is still with us if we hold onto the experiences that suggest that there is a magnitude which is physical distance (space) and a magnitude which is the physical duration (time).

Russell's solution to Zeno's paradox is radical. He maintains that we should abandon our empirically informed ideas of *motion,* and *space, change* and *time* in favor of purely mathematical reconstructions. It is not often realized that Russell held this position as early as 1901. But the point comes through well enough in the following passage:

> Weierstrass, by strictly banishing from mathematics the infinitesimal, has at last shown that we live in an unchanging world, and that the arrow in its flight is truly at rest. Zeno's only error lay in inferring (if he did infer) that, because there is no such thing as a state of change, therefore the world is in the same state at any one time as at any other. This is a consequence which by no means follows; and in this respect, the German mathematician is more constructive than the ingenious Greek. (Russell 1901, 63)

In Russell's view, the ordinary notions of *matter*, *space* and *change* are bundles of confusions informed by misleading empirical experiences with rocks and pebbles. These are to be replaced by constructions made possible by the mathematics of the infinite. For the most part, Russell's mathematical reconstructions rely on Newtonian laws. Russell was willing to entertain Leibniz's notion that space is relational (not absolute), but in 1901 he never fathomed that time itself might be relative. Indeed, Russell's purely mathematical approach wholly depends on the current physical theory of his day to set forth the "laws" to be realized by the eliminativistic reconstructions. But in any case, Russell's reconstruction of these notion is designed so that the new mathematical definitions of *limits* and *continuity* apply to them. Thus, for Russell there is no gap between the mathematical constructions and the physical world. Zeno's paradoxes are solved in virtue of the elimination of the ordinary notions of *matter*, *space* and *time* and *change*.

3. There is no matter

It is rarely appreciated that as early as *Principles* Russell was advocating eliminativistic reconstructions of ordinary notions of *matter*. Let me quote at length his attempt at a reconstruction. He writes:

> We may sum up the nature of matter as follows. *Material unit* is a class-concept, applicable to whatever has the following characteristics: (1) A simple material unit occupies a spatial point at any moment; two units cannot occupy the same point at the same moment. (2) Every material unit persists through time; its positions in space at any two moments may be the same or different; but if different, the position at times intermediate between the two chosen must form a continuous series. (3) Two material units differ in the same immediate manner as two points or two colours; they agree in having the relation of inclusion in a class to the gen-

> eral concept *matter*, or rather to the general concept *material unit*. Matter itself seems to be a collective name for all pieces of matter, as space for all points and time for all instants. It is this peculiar relation to space and time which distinguishes matter from other qualities, and not any logical difference such as that of subject and predicate, or substance and attribute.
>
> We can now attempt an abstract logical statement of what rational Dynamics requires its matter to be. In the first place, time and space may be replaced by a one-dimensional and a *n*-dimensional series respectively. Next, it is plain that the only relevant function of a material point is to establish a correlation between all moments of time and some points of space, and that this correlation is many-one. So as soon as the correlation is given, the actual material point ceases to have any importance. Thus we may replace a material point by a many-one relation whose domain is a certain one-dimensional series, and whose converse domain is contained in a certain three-dimensional series. To obtain a material universe, so far as kinematical considerations go, we have only to consider a class of such relations subject to the condition that the logical product of any two relations of the class is to be null. This condition issues impenetrability. If we add that the one-dimensional and the three-dimensional series are to be both continuous, and that each many-one relation is to define a continuous function, we have all the kinematical conditions for a system of material particles, generalized and expressed in terms of logical constants. (Russell 1903, 468)

Of course, Russell's conception of physical *space* in *Principles* differs significantly from the conception he offers in neutral monism of the 1920's. This is not surprising since the latter conception, and not the former, adopts Einstein's theory of relativity. Neutral monism (with neutral stuff as minimal events of space-time) is a four-dimensionalist model of space-time. But in both periods of Russell's work, it is the purely mathematical reconstructions that carry the day.

In his *A History of Western Philosophy*, Russell offers a wonderful assortment of examples of logical analyses that illustrate his new scientific/mathematical conception of philosophy. "Physics," Russell tells us, "as well as mathematics, has supplied material for the philosophy of philosophical analysis. What is important to the philosopher in the theory of relativity is the substitution of space-time for space and time." In 1915 Einstein's general theory of relativity bested Newton's theory of gravitation in finally providing an adequate explanation of the perihelion of the planet

Mercury. By 1919 an expedition led by Arthur Eddington confirmed general relativity's prediction for the deflection of starlight by the sun, making Einstein famous. By 1927, Russell's *Analysis of Matter* appeared and it shows that Russell had fully subsumed the theory into his philosophy of space and time.

Experience with sticks makes it seem that the notion of "physical length" is intuitively clear as an intrinsic, non-relational, property of a body. Motion seems to require the accumulation of physical lengths. Of course, scientists knew that temperatures and pressures deform sticks, even those of iron and steel and concrete. But fixing the parameters, this did not jeopardize the traditional concept of physical length as a property. Einstein maintains that physical length and time are properties relative to inertial frames. Protagoras said that "man is the measure of all things (physical)." Einstein says that the measure of all things physical is the invariant behavior of electromagnetic energy (light).

Russell's discussion in *The A B C of Relativity* (1925) remains today one of the most accessible explications of Einstein's theory, maintaining (contrary to a host of popular discussions) that the theory is about the world itself and not merely about practices of measurements by observers. Russell sees clearly that it is a theory of the relativity of physical length, not a theory that lengths contract under accelerations approaching light speed. Special relativity, and its symmetries of uniform motions and seeming "distance contractions" and "time dilations", is merely a heuristic device. The correct theory (the only theory of relativity) is the General Theory of Relativity. If light is emitted in the center of a square box at t_1 and if relative to inertial frame A the box is moving as in the picture (below), then at t_2 the light propagates so as to contact the side of the box rushing toward it and producing event e_1 before event e_2. It is as if the light propagates as a transverse wave of electro-magnetism in a luminiferous *aether* which permeates all objects and is unaffected by their motions.

On the other hand, relative to the inertial frame B of the box itself, the light propagates as if particles would without a luminiferous *aether*. It contacts both sides of the box at the same time. Event e_1 and e_2 are simultaneous. This may at first seem contradictory, but it is well supported by the famous Michelson-Morley experiment of 1887 (and has been corroborated many times since). Physicists Lorentz and FitzGerald independently proposed length contractions to solve it. In contrast, Einstein offered a conceptual revolution: physical lengths and physical times are not invariant properties independent of inertial frame. The propagation of light is the invariant.

Progagation of light is invariant in all inertial frames.

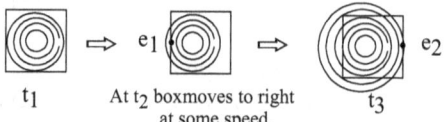

t_1 At t_2 boxmoves to right at some speed. t_3

In inertial frame A, the light does not take on the inertia of the box. Event e_1 occurs before event e_2.

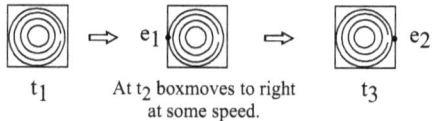

t_1 At t_2 boxmoves to right at some speed. t_3

In inertial frame B, the light propagates in the intertial frame of the box. Event e_1 is simultaneous with event e_2.

Russell regards Einstein as offering a philosophical transformation of the traditional physical concepts of "space" and "time" into "space-time." Ongoing events of light radiation are invariant in all inertial frames. Thus distance and time will be determined by the behavior of light, and not otherwise. That is, we cannot pick a physical oscillation (clock of some sort) and an object, and, pretending the object is a rigid bit of matter (having a determinate length and mass that is invariant in all inertial frames), use it to understand the behavior of light. It is light's behavior that determines length (space) and time itself. The essence of the theory is the Principle of Relativity. Let me state it as follows: In any inertial frame, an object is always at the center of its own light cone. The light cone is produced by light propagating in all directions from the source through space-time. Hence all events on the lip of the light cone are, by definition, simultaneous. The Principle of Relativity is impossible if space (physical distance, length of a bit of matter) and time are absolute (invariant of inertial frame) as Newton thought. But if an object, moving or otherwise, is always at the center of its own light cone, then space gives way to space-time (space and time are connected) and are not invariant of one's inertial frame. We can calculate relative spatial lengths and time simultaneity in inertial frames by imagining space-time as a plane intersecting a light cone, as illustrated in the picture.

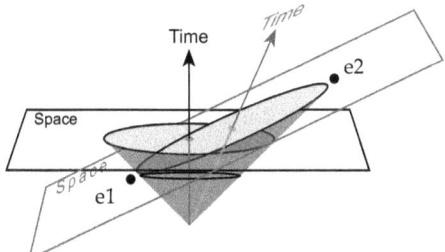

All events along the circular lip of the light cone in the horizontal space-time plane are simultaneous, as are all events along the lip of the circular light cone in the tilted space-time plane. The light emitter in the tilted plane is moving at some speed away (to the right) with respect to the light emitter in the horizontal plane. As we can see, in order for horizontal and tilted light emitters to stay in the center of their respective light cones, events that are simultaneous-horizontal with e_1 occur before events that are simultaneous-tilted with e_2. Matter, as a rigid body with fixed length, and in absolute time, has to be abandoned. Indeed, mass itself (as the propensity of an object to resist changes in its inertia) must be relative to inertial frame as well and not invariant as in Newton's physics.

Russell's willingness to abandon ordinary notions of *matter*, *space* and *time* does not end with relativity. Turning to quantum theory, Russell writes: "I suspect that it will demand even more radical departures from the traditional doctrine of space and time than those demanded by the theory of relativity."[15] Russell approvingly cites the new work in quantum theory as important for the logical analysis of *matter*. And he counts Einstein's thesis that "simultaneity" has to be *defined* as just such a separation of the logical from the empirical (Russell 1945, 832).

Just as Einstein reconceptualizes Maxwell's laws for the propagation of light without the luminiferous aether, Russell's logical atomism abandons the non-logical ontological necessities of former speculative philosophies, offering a re-conceptualization and *logical* reconstruction (where possible) of the laws of the earlier theory. Many successes of the earlier theories are retained by the reconstruction. Retention, however, is always partial. Some processes and mechanisms of earlier theories are treated as

---

[15] *Op Cit.*, 833.

flotsam. The supplanting tradition may come to regard the terms of the earlier theories as non-referential, or regard earlier ontologies as idle wheels that serve no explanatory purpose.

In 1919, emboldened by Eddington's interpretation of Einstein's theory relativity, Russell adopted a four-dimensionalist version of neutral monism. Russell's neutral monism merges a four-dimensionalism with respect to matter (continuants persisting in time) with James's thesis that consciousness is a process: Conscious selves persisting through time, no less than physical continuants, are series of space-time events. It is worth quoting at length from the opening of *The Analysis of Mind*:

> On the one hand, many psychologists, especially those of the behaviorist school, tend to adopt what is essentially a materialistic position, as a matter of method, if not metaphysics. They make psychology increasingly dependent on physiology and external observations, and tend to think of matter as something much more solid and indubitable than mind. Meanwhile, the physicists, especially Einstein and other exponents of the theory of relativity, have been making "matter" less and less material. Their world consists of "events," from which "matter" is derived by a logical construction. Whoever reads, for example, Professor Eddington's *Space, Time and Gravitation* (Cambridge University Press, 1920), will see that an old-fashioned materialism can receive no support from modern physics. I think that what has permanent value in the outlook of the behaviorists is the feeling that physics is the most fundamental science at present in existence. But this position cannot be called materialistic, if, as seems to be the case, physics does not assume the existence of matter. The view that seems to me to reconcile the materialistic tendency of psychology with the anti-materialistic tendency of physics is the view of William James and the American new realists, according to which the "stuff" of the world is neither mental nor material, but a "neutral stuff," out of which both are constructed. (Russell 1921, 5)

Succinctly put, Russell says that the old distinction between soul and body (mind and matter) has evaporated quite as much because "matter" has lost its old solidity as because 'mind' has lost its spirituality (Russell 1935a, 133). Russell accepts Eddington's interpretation that relativity eliminates the traditional notion of "matter" in favor of series of events in space-time. In 1926 Russell even retrofitted his earlier work *Our Knowledge of the External World as a Field for Scientific Method in Philosophy* (1914) with the new view, replacing physical sense-data, which he had formerly taken to be the basis of the construction, by physical events in space-time.

Relativity brought to the fore how imperative it is to free physics from the empirically informed intuitions of physical *space, matter, change* and *time*. Is the notion of exactly half of a stick intelligible? And is there any physical sense in which sticks have infinitely many parts? In relativity theory, the answer to such questions depends on whether physical events overlap sufficiently so that physical correlations (given by light interactions) of parts of those events are fine enough to be modeled by the *dense* structure of rational numbers. (The rationals are dense in the sense that for any two, there is one in between them.) Physical correlations by light interactions do seem at least to enable measuring a given distance by the ratios. For example, by rotation on a lathe, a stick can be scribed with marks that are, given the needed practical applications, reasonably close to being equidistant from one another. They are not, of course, "equidistant" from one another in the mathematical sense. But no practical problems arise unless some new needs force themselves upon us. If they do, we can improve the scribed marks on the stick by other, more refined physical processes. By appeal to such physical correlations, it is intuitive that it is physically possible in principle to improve our accuracy of measurement to any degree modeled by the structure of the ratios.

Can we improve our physical measuring device to a degree that supports the application of the real numbers? In applying the differential calculus to model change of position of a moving object over time, we have a ratio of time to position. This ratio is made by correlations between parts of events which overlap. The events that inform the construction of the notion of a velocity at a spatial point and instant of time are *ordinary* ongoing processes of everyday experience—processes we naturally describe by employing reference to ordinary material continuants: the flowing of the Mississippi into the Louisiana delta, the melting and freezing of the glaciers, light reaching the earth from the Sun, the expansion of the universe. These are quite ordinary overlapping events whose parts are naturally supposed to be ordinary material objects persisting in time. Consider the case where one event is the motion of the hands of a clock and the other is the motion of the physical continuant across the field whose ground is scribed. When the clock's second hand was at x the object was at hash mark y, etc. The definition of a velocity at a temporal moment is made possible only by appeal to the limit of the function characterizing the changes in the ratio of distance to time in the overlapping correlations of the events of the motion of the object. (It is now common to illustrate this in analytic geometry as an application of the problem of finding the tangent to a curve "at a point.") The notion of instantaneous velocity is the derivative (a limit) of a function. On Weierstrass's reconceptualization of the notion of the *limit* of a function, the limit is not an infinitesimal. Weierstrass's notion of the limit requires a neighborhood, some of whose ratios of distance to time are future to the point to be constructed and some whose ratio of distance to time are before it. The notion of the leading point of an *ongoing* motion cannot be defined. It is possible to speak (by Weierstrass's construction) of an event (e.g., of having a velocity) *at an instant* only in virtue of the presence of overlapping ordinary events involving physical continuants.

In Russell's view, it is a contingent empirical matter whether the physical correlations by light interactions of ordinary overlapping events are fine enough to enable the construction of talk of "temporal instants." In his essay "On Order in Time" (1936) Russell sets out the parameters needed for the construction and notes that "[…] the existence of instants requires hypotheses which there is no reason to suppose true—a fact which may be not without importance in physics" (Russell 1936, 347). Thus Russell leaves open the matter of whether there are physical processes by means of which distances can have real number measures.

Russell's book *Our Knowledge of the External World as a Field for Scientific Method in Philosophy* (1914) was revised several times in light of the new physics. Sense-data (1912-1915), as the physical complexes presented to the sense organs and sensed by a 'self,' are abandoned. Sensation is a physical-neural process of events. There is no matter (rigid bodies enduring in space and time); the laws of physics governing matter are reconstructed so as to be realized by series of physical events ("transient particulars"). There is no mind (self enduring in time); the (largely behavioristic) laws of psychology governing minds are reconstructed so as to be realized by series of physical events ("transient particulars"). Some transient particulars may occur only in series that are minds (e.g. "images"); some may occur only in series that are matter. But the transient particulars are space-time events (the neutral "stuff" of both). Russell writes:

> The truth is, of course, that mind and matter are, alike, illusions. Physicists, who study matter, discover this fact about matter; psychologists, who study mind, discover this fact about mind. But each remains convinced that the other's subject of study must have some solidity. What I wish to do in this essay is to restate

> the relations of mind and brain in terms not implying the existence of either. (Russell 1956, 145)

Russell writes:

> Matter as it appears to common sense, and as it has until recently appeared in physics, must be given up. The old idea of matter was connected with the idea of "substance," and this, in turn, with a view of time that the theory of relativity shows to be untenable. [...] Events which can be regarded as all in one place, or all parts of the history of one piece of matter, still have a definite time-order. So do events in different places [...] if light can travel from the place of the one to the place of the other so as to reach the other place before the second event [...]
>
> What do we mean by 'piece of matter' [...] we do not mean something that preserves a simple identity throughout its history, nor do we mean something hard and solid, nor even a hypothetical thing-in-itself known only though its effects. We mean the 'effects' themselves, only that we no longer invoke an unknowable cause for them. When energy radiates from a centre, we can describe the laws of its radiation conveniently by imagining something in the centre, [...] All this, however, is only a convenient way of describing what happens elsewhere, namely the radiation of energy away from the centre. As to what goes on in this centre, if anything, physics is silent.
> [...] Materialism as a philosophy becomes hardly tenable in view of this evaporation of matter. (Russell 1927, 158)

We have seen as well that in *Principles* Russell adopted the view that *matter*, in the ordinary sense, does not exist. But he was not then a four-dimensionalist. Russell's views in the 1920's, however, show him to unequivocally accept four-dimensionalism. He says that, properly reconstructed, material objects are tubes (series) of temporal parts or stages spread out in time, with their temporal parts or stages occurring at each minimal amount of time. Russell writes:

> We will define a set of compresent events as a "minimal region". We find that minimal regions form a four-dimensional manifold, and that, by a little logical manipulation, we can construct from them the manifold of space-time that physics requires. We find also that, from a number of different minimal regions, we can often pick out a set of events, one from each, which are closely similar when they come from neighbouring regions, and vary one re-

> gion to another according to discoverable laws. These are the laws of the propagation of light, sound etc. We find also that certain regions in space-time have quite peculiar properties; those are the regions which are said to be occupied by "matter". Such regions can be collected, by means of the laws of physics, into tracks or tubes, very much more extended in one dimension of space-time than in the other three. Such a tube constituted the "history" of a piece of matter; from the point of view of the piece of matter itself, the dimension in which it is most extended can be called "time," but it is only the private time of that piece of matter, because it does not correspond exactly with the dimension in which another piece of matter is most extended. Not only is space-time very peculiar with a piece of matter, but it is also rather peculiar in its neighborhood, growing less so as the spatio-temporal distance grows greater; the law of this peculiarity is the law of gravitation. (Russell 1924, 342)

Russell is advocating a form of physicalism of space-time events (physical transient particulars).

This form of neutral monism is not a reductive identification of matter and minds with entities (tubes) of space-time. It is the straightforward non-existence of matter and mind as single objects. This kind of eliminativism, as opposed to reductive identification, is not easy to understand. Russell's eliminativism comes from his constructive work in mathematics. It is there that we find our best understanding of it. Make no mistake about it, the neutral stuff of Russell's neutral monism is not neutral. And radical as it may seem, Russell is offering an elimination, not a reductive identification. There is no mind and there is no matter.

4. *Eternalism's block universe or the growing block or presentism*

Russell clearly holds a form of the B-theory and a four-dimensionalism which maintains that the solution of paradoxes involves abandoning ordinary notions in favor of eliminativistic reconstructions of *space* and *time* and *matter*. But in saying this we must hasten to add that Russell's eliminativism and constructivism are central to any proper understanding of his views on the metaphysics of time. The two positions are not always at peace with one another. We see this when we ask about the nature of untensed events. If the notion of an untensed event is itself a logical fiction, then we can scarcely regard Russell as a straightforward B-theorist and four-dimensionalist. And yet, we find Russell suggesting that events are logical fictions. He writes:

> Most people have discarded 'time' as something distinct from temporal succession, but they have not discarded 'event.' An 'event' is supposed to occupy some continuous portion of space-time, at the end of which it ceases, and cannot recur. It is clear that a quality, or complex of qualities, may recur; therefore an 'event,' if non-recurrence is logically necessary, is not a bundle of qualities. [...].The view that I am suggesting is that an 'event' may be defined as a complete bundle of compresent qualities, i.e., a bundle having the two properties (*a*) that all the qualities in the bundle are compresent, and (*b*) that nothing outside the bundle is compresent with every member of the bundle. I assume that as a matter of empirical fact, no event recurs; that is to say, if *a* and *b* are events, and *a* is earlier than *b*, there is some qualitative difference between *a* and *b*. For preferring this theory to one which makes an event indefinable, there are all the reasons commonly alleged against substance. (Russell 1948, 83)

"Compresence" suggests present at the same time, and thus it seems viciously circular to imagine a B-theory in which untensed events have temporal orderings, and yet turn around and maintain that events are, themselves, logical fictions.

We saw that in "On Order in Time" Russell explored the requirements for constructing temporal instants, showing that the physical requirements for continuity may not be met. Nothing in four-dimensionalism or the B-theory requires that temporal moments be adopted as indefinable. But focusing on times as instants to which events stand in precedence relations hides the issue. The problem is not with times as relata. The problem is with events themselves—that is, the untensed events that have endpoints (or beginning points). The very untensed events (events with beginning or endpoints) out of which B-theorists compose series seem to be logical fictions in Russell's metaphysics of time. Such events are relations of other events. This follows because the beginning or end points of events are logical fictions that live in relations between events that overlap. To speak of an event as having ended, it has to have an end point. And thus any such event, along with its end point is a construction (a relation among ongoing events). Any event with a beginning or end point is a relation among overlapping events. These overlapping events may well themselves have beginning or end points, but if so, they too live in relations of yet other overlapping events. The construction has to stop somewhere, else it is viciously circular.

I am not certain if Russell ever fully appreciated the tension between his constructivism and his B-theory of time. We see it emerging after "On Order in Time" and only faintly in *Human Knowledge*, which is the very last of his works in metaphysics. Events with a beginning or an end are relations on other events. They are relations among overlapping events. Thus we see that the notion of an *ongoing* event is primitive and unanalyzable. An ongoing event is, in a sense, a tensed event. Untensed events (those with endpoints) are constructed in terms of relations of overlapping ongoing events. The analogy with the calculus is helpful. There are no infinitesimal volumes which, taken together, constitute volume. There are no instantaneous velocities which, through integration, constitute velocity. We speak legitimately of matter being constituted by series of untensed events of a fixed duration only in the way we speak of the *composition* of the integral. The construction of minimal intervals of time, minimal enduring events, in "On Order in Time" logically requires that there be neighboring overlapping events. There must be events with parts that are future to the given constructed event (with an end point), and events whose parts are past with respect to it. If this construction is not to be viciously circular, we are driven to the primitive undefinable notion of an *ongoing* event. All such constructed events live as relations among ongoing events.

Perhaps the expansion of the universe itself is the best example of a genuinely ongoing event—an event with no beginning point and no end point. But in order to provide the foundation of a Russellian construction of untensed events (with beginning and end points) there must be a great many such events. Currently there is little consensus among physicists about the events taking place near the origins of the universe. Most theories are highly speculative and general relativity and quantum theory have not been fully reconciled. It is thought, however, that between 10^{-6} seconds and 1 second after the Big Bang, protons and neutrons form and neutrinos decouple and begin traveling freely in space creating a cosmic background radiation (the microwave radiation background came later). Speculative theories attempt to imagine events much earlier, events with a bound of 10^{-43} seconds after the Big Bang when the four fundamental "forces" (electromagnetism, gravitation, weak interaction and strong interaction) all have the same strength and might be united. Physical cosmologists speculate on quantum gravitation, string theory and the like, hoping that some such theory might best describe events during this epoch. But however this comes out, Russellians can allow the characterization of events with beginnings in such an early epoch only by appeal (ultimately) to events that are ongoing. The notion of an event whose beginning point is the origin of the universe is, on the present view, an unintelligible notion. As we have seen, events overlapping in a neighborhood before and after such a "beginning instant" are required for the construction. But, *ex hypothesis*, there are no such overlapping events for the origin (beginning) of the universe.

Of course, physicists working on cosmology endeavor to find evidence of the beginnings of earlier and earlier events. Their endeavors rely upon events with beginning points (as well as events with beginning and end points) already assumed. This working backwards is not undermined by the thesis that *ongoing* events are ultimately the foundation of all events. Physicists work with plenty of ongoing events—even if some of the events they thought were genuinely ongoing turn out, after investigation, to have had beginnings (e.g., the origins of hydrogen, separation of weak nuclear force, etc.).

Modern B-theorists and four-dimensionalists are no friends of construction of untensed events and the primacy of a notion of an *ongoing* event.[16] All events are on a par, and ontologically tenseless. It is a form of eternalism of events—the "Block Universe." In contrast, Russell offers constructions of moments of time and minimal enduring events. They are logical fictions. Like the ordinary four-dimensionalists, he maintains that there is no "matter" (no traditional rigid bodes that endurantist materialists accept). Matter is a construction from events. But unlike the Eternalist, Russell's construction of matter is ultimately from ongoing events. The result offers a new position in the debates over the nature of time.

Recall the dispute between the "Growing Block" theory of the nature of time and the Eternalist or "Block" theory of the nature of time. According to the Growing Block theory, there is a leading edge of becoming, and events future to this leading edge are unreal. If Russell (tacitly) accepts a primitive unanalyzable notion of events which are *ongoing,* he cannot be placed in the camp with the Eternalists. But neither can he be placed with the Growing Block theory of the nature of time. The reason is that, due to Russell's constructivism, no sense can be made of the notion of *becoming*. No sense can be made of it because the absence of the neighborhood entails that the notion of a minimal enduring event or a minimal leading edge of the process of *becoming* cannot be constructed. The construction of an

---

[16] See Sider 2001.

event, an interval with an endpoint, requires that there be events that have parts overlapping with it that are *future* to every event with an end point in the interval. To say that an event e is untensed is to say that it has a beginning point and an endpoint. But if one accepts Russell's constructivism, where there is an event e with an end point, there must necessarily be events future to e. Thus, not all events are untensed. Some are ongoing and relations among these are, ultimately, the foundation of the construction of the logical fictions that are untensed events. Though the construction accepts *ongoing* as primitive, *becoming* is a pseudo-concept. It cannot be constructed.

Presentism is also ruled out by Russell's constructivism. Presentism rejects Eternalism's block view of the nature of time with its ontology of events ordered serially and so that none of them have a special privileged status of having the property of *occurring now* (being present). Obviously, Caesar does not exist any longer. Obviously, the event of *Cicero's denouncing Catiline* is not occurring any longer. It ended. And with its end it ceased to exist. Were it to exist presently, then its constituent Caesar would also exist presently—or so says the Presentist. Russell regards that as a confusion. Caesar (both his mind and his body) are themselves series of events. These events have ended; they had beginning points and end points. It is in this sense that Caesar no longer exists. But the events that constitute him do exist. The events exist, and presently exist, though they are no longer unfolding. Perhaps a Presentist can find some consolation in the thesis that these events are relations of ongoing (unfolding) events. But a Russellian robust sense of reality does not lead to Presentism. Quite the contrary, we now see that a robust constructivism regards untensed events as logical fictions (relations of ongoing events), and this reveals that the Presentist conception of a present event, whether a specious present (with beginning and/or end point) or an instant, is incoherent.

The lesson for Russellians is this: ongoing events are primitive and unanalyzable. They are the foundational "stuff" of the construction, not only of matter and mind but also of untensed events. Eternalism, Growing Block, and Presentism do not exhaust the positions on the nature of time. There is yet Russell on the metaphysics of *Time*.

REFERENCES

Bourne, C. (2006). *A Future for Presentism*. Oxford: Clarendon Press.

Coffa, A. (1977). Carnap's Sprachanschauung circa 1932. *Philosophy of Science Association*, 205-241.

McTaggart, J. (1908). The Unreality of Time. *Mind: A Quarterly Review of Psychology and Philosophy*, 17, 456-73.

McTaggart, J. (1921). *The Nature of Existence*, Volumes 1 & 2. Cambridge: Cambridge University Press.

Prior, A. N. (1968). Now. *Noûs*, 2, 101-119.

Quine, W. V. O. (1948). On What There Is. In *From a Logical Point of View*, Cambridge: Harvard University Press, 1964, 1-19.

Quine, W. V. O. (1951). Carnap's Views on Ontology. *The Ways of Paradox and Other Essays*, Cambridge: Harvard University Press, 1976, 203-211.

Ramsey, F. P. (1931). Philosophy. *The Foundations of Mathematics and Other Essays by Frank Plumpton Ramsey*, ed. R. B. Braithwaite, New York: Harcourt, Brace and Co., 263-269.

Russell, B. (1901). Mathematics and the Metaphysicians. *Mysticism and Logic and Other Essays*, New Jersey: Barnes and Noble, 1976, 74-96.

Russell, B. (1903). *The Principles of Mathematics*. Cambridge: Cambridge University Press.

Russell, B. (1914). *Our Knowledge of the External World as a Field for Scientific Philosophy*. London: Allen & Unwin.

Russell, B. (1919). *Introduction to Mathematical Philosophy*. London: Allen & Unwin, 1953.

Russell, B. (1921). *The Analysis of Mind*. New York: Allen & Unwin, 1961.

Russell, B. (1924). Logical Atomism. *Essays in Analysis by Bertrand Russell: 1901-1950*, ed. R. Marsh, London: Allen & Unwin, 1977.

Russell, B. (1925). *The ABC of Relativity*. New York: Harper & Brothers.

Russell, B. (1935a). *Religion and Science*. Oxford: Oxford University Press.

Russell, B. (1935b). The Limits of Empiricism. *Proceedings of the Aristotelian Society*, New Series, Vol. 36 (1935-1936), 131-150.

Russell, B. (1936). On Order in Time. *Essays in Analysis by Bertrand Russell: 1901-1950*, ed. R. Marsh, London: Allen & Unwin, 1977.

Russell, B. (1945). The Philosophy of Logical Analysis. *A History of Western Philosophy* (Chapter 31), New York: Simon and Schuster.

Russell, B. (1948). *Human Knowledge: Its Scope and Limits*. New York: Simon & Schuster.

Russell, B. (1956). *Portraits from Memory and Other Essays by Bertrand Russell*. New York: Simon & Schuster.

Russell, B. (1968). *The Art of Philosophizing*. New York: The Philosophical Library.

Sainsbury, R. M. (2009). *Paradoxes*. Cambridge: Cambridge University Press.

Sider, T. (2001). *Four-Dimensionalism*. Oxford: Clarendon Press.

Thomson, J. (1954). Tasks and Super-Tasks. *Zeno's Paradoxes*, ed. W. Salmon, New York: The Bobbs-Merrill Co., 1970, 89-102.

WEAK DISCERNIBILITY AND THE IDENTITY OF SPACETIME POINTS

DENNIS DIEKS
Utrecht University
d.dieks@uu.nl

ABSTRACT. In this article we investigate putative counterexamples to Leibniz's principle of the Identity of Indiscernibles. In particular, we look at the status of spacetime points: although these all possess exactly the same properties in symmetrical spacetimes and thus seem indiscernible, there are certainly more than one of them. However, we shall defend Leibniz's principle, even for such highly symmetrical cases. Part of our strategy will be to invoke the notion of "weak discernibility", as proposed in the recent literature. Weakly discernible objects share all their properties (both monadic and relational) but stand in *irreflexive* relations to one another—this irreflexivity makes it possible to show that Leibniz's principle implies their numerical diversity. However, we shall argue that this notion of weak discernibility only serves its purpose if it is supplemented by a criterion of physical meaningfulness of the relations and relata in question; and that in the final analysis weak discernibility only helps out when it can be seen as a degenerate case of strong, absolute discernibility. The case of spacetime points is an example of such a situation.

1. Introduction

It is an uncontroversial principle of modern physical science that the formulation of theories and hypotheses should be guided by empirical data; physical science should eschew concepts that are superfluous from an empirical point of view. As a case in point, modern science is averse to the use of the notion of *haecceity*, "primitive thisness", in order to individuate objects. It may be acceptable to scholastic philosophy, so the thought goes, to say that objects differ from each other by virtue of some non-empirical internal principle that bestows individuality on them, but objects as conceived in physical science derive their identity from distinguishing physical features, which are in principle open to empirical investigation. Physical science is thus sympathetic to Leibniz's principle of the identity of indiscernibles, which says that entities have to be discernible in order to be different, with discernibility interpreted in terms of physical concepts. Indeed, if there were no physical, empirical distinctions at all in a candidate case of numerical diversity, then why should we even start supposing that we are facing distinct entities?

Natural and plausible as these ideas seem, their application is not without difficulties. Even situations that are easy to imagine on the basis of everyday experience seem capable of sometimes violating Leibniz's principle—the prime example being symmetrical configurations of objects with the same material properties, in an otherwise empty universe (e.g., "Black's spheres", to be discussed below). The situation becomes worse in fundamental physical theories like quantum mechanics and relativity. In the latter case, the problem is that spacetime points in symmetrical spacetimes (e.g., Minkowski spacetime) have all their geometrical properties in common, so that Leibniz's principle would appear to lead to the absurd conclusion that there can be only *one* point in such spacetimes.

In this article we shall further explain these difficulties and then go on to defend Leibniz's principle, even for the highly symmetrical cases just mentioned. Part of our strategy will be to invoke the notion of "weak discernibility", as proposed in the recent literature. Weakly discernible objects share all their properties (both monadic and relational ones) but stand in *irreflexive* relations to each other—this irreflexivity makes it possible to show that Leibniz's principle implies that there are more than one of them. However, we shall argue that the application of this notion of weak discernibility only achieves its aim if it is supplemented by a criterion of physical meaningfulness of the relations and relata in question; and that in the final analysis weak discernibility only helps out in cases in which it can be seen as a degenerate case of strong, absolute discernibility.

2. The identity problem

It is part and parcel both of ordinary life and the sciences to think of the world in terms of different individual entities. Also both in ordinary life and the sciences, it seems *prima facie* obvious that empirical distinctions should accompany these differences in individuality—indeed, how could we have arrived at the concept of an object in the first place if not on the basis of such empirical differences?

In physics we accordingly expect that different individuals (different particles, for example) are characterized by different values of at least some physical quantities. By "physical quantities" we mean those general (qualitative) predicates that figure in physical laws and are thus relevant for empirical predictions—predicates like "mass", "charge", "velocity", etc. We could of course posit the individuality of objects by *fiat*, via the introduction of a notion of "primitive thisness" or "*haecceity*", so that each object comes to possess its own "thisness" that distinguishes it from all other objects. But if these new concepts do not go together with ordinary, empirically accessible physical differences they cannot do any work in physical theory, which would make their introduction undesirable.

These expectations and intuitions are in agreement with the ideas behind Leibniz's *principle of the identity of indiscernibles* (PII). Roughly speaking, this principle says that two putative objects having all their properties in common are actually one and the same object. In order to be in accordance with what was just said, we have to restrict the domain of the properties considered in PII so as to include only physical quantities. If we take these quantities to include absolute position, as is reasonable within Newtonian mechanics with its absolute space, PII is fully satisfied by classical particles (French and Krause 2006, ch. 2). Indeed, classical particles are always taken to be impenetrable so that they cannot occupy the exact same position in absolute space. Even if two Newtonian particles possess exactly the same mass, charge, and other physical attributes, they must still differ in where they are—and this is sufficient to ground their individuality in a physically acceptable way.

However, on second thought complications arise, even for classical physics. The status of absolute space has been a bone of contention in the history of physics: it has been proposed repeatedly that space in itself is not a viable physical concept, and that we should rather think along the lines of a Leibnizean relationism. According to the relationist position, in a classical particle universe only the particles themselves exist, without being embedded in an independently existing space. The essential relationist idea is that the particles can nevertheless have spatial relations with respect to each other—these are possessed *directly* by the particles and do not derive from pre-existing relations between the spatial points occupied by the particles.

Now, we do not need to take sides in the relationist-absolutist debate in order to acknowledge that relationism is a conceptual possibility. It seems natural to accept the conceptual consistency of considering two (or more) particles that have distances and orientations relative to each other without being pinned down in an absolute background space. But this conceptual possibility spells trouble for Leibniz's PII as a universally valid principle.

Indeed, even staying within the domain of everyday experience and without entering physical theory proper, it now seems possible to construct counterexamples to PII. These are cases in which it seems intuitively obvious that there are several individuals, but in which there are no distinguishing qualitative characteristics, so that the individuals are indiscernible. One famous case, proposed by Max Black (1952), is that of two perfect spheres of identical chemical composition and at a mutual distance of two miles in a relational universe (so there is no absolute background space). Another example is provided by a relational universe consisting of two hands that are each other's mirror images (Kant's enantiomorphic hands). The essential feature of these cases is that the objects have all their properties in common, both as far as monadic and as far as relational predicates are concerned. Thus, both spheres in Black's example have the same material characteristics, and both are at two miles from a sphere. Similarly, Kant's hands each have the same internal geometric properties and both are mirror images of a hand. So these cases appear to demonstrate that already in everyday life we think in terms of notions of "object" and "individuality" that are independent of the presence of distinguishing qualitative differences—in violation of PII.

The situation becomes even more urgent when we consider modern physical theory. A notorious obstacle for PII comes from quantum mechanics, in which the notion of "identical particles" plays a major role. These are particles of the same kind, i.e., with the same intrinsic properties (like mass, charge, spin); e.g., electrons, protons or neutrons. It is a basic principle of quantum mechanics that the state of a collection of such particles must be completely symmetrical (in the case of bosons) or antisymmetrical (fermions) with respect to the particle indices occurring in it. This symmetrization postulate implies that all one-particle pure states represented in the total state describing a collection of identical particles occur symmetrically in it: exchanging two such one-particle states in the total state leaves the total state invariant (apart from a physically insignificant change of sign in the case of fermions). It does therefore not matter which particle index we associate with which state. It follows from this symmetry that any property or relation that can be attributed, on the basis of the total quantum state, to any one particle has to be attributed to each of the other particles as well.[1] Again, we are facing an apparent violation of PII.

---

[1] Another way of expressing this is that the one-particle "states" defined by "partial tracing" are the same for all indices.

The possible counterexamples to PII that will concern us first of all in this article are of a geometrical nature. To start with, consider the (infinite) Euclidean plane. This is a highly symmetrical geometrical object: it is invariant under translations, rotations and reflections. It follows that there are no privileged points in the Euclidean plane—indeed, symmetry transformations can transform any given point into any other, leaving all geometrical properties of these points and all mutual geometrical relations the same. This entails that each and every point in the plane possesses exactly the same geometrical status: the plane looks exactly the same viewed from whatever point. Clearly then, the points in the Euclidean plane fail to be discernible on the basis of their geometrical properties. If we assume that there are no haecceities that transcend these geometrical properties, application of Leibniz's principle PII then appears to lead us to the (absurd) conclusion that there is only *one* point in the Euclidean plane.

The same argument can be given a more physical turn if we apply it to Newton's absolute space. However, let us immediately go on to special relativity theory and consider empty Minkowski spacetime. Special relativity posits four-dimensional Minkowski spacetime as the fixed geometrical background against which all physical processes evolve. Minkowski spacetime is endowed with a distance function: between any two points in it a definite distance is defined. The situation is more complicated, however, than in the Newtonian or Euclidean case: the special relativistic distance function can assume both positive and negative values, and moreover there are non-coinciding points between which the Minkowski distance is zero. Relative to any given point in Minkowski spacetime the total spacetime can accordingly be divided into three parts: all points that have a positive spacetime distance with respect to the chosen point (points with time-like separation); points at zero distance (points with light-like separation); and points that possess a negative distance (points with space-like separation).[2] The points with time-like separation lie in the interior of the lightcone that can be drawn from our fiducial point; the points with space-like separation are outside of the lightcone; the lightcone itself consists of all points whose distances to the fiducial point vanish. Because of this more involved metrical structure the symmetries of Minkowski spacetime assume a more complicated form than those of Euclidean space, but it remains nevertheless true that there are no privileged points. Minkowski spacetime looks exactly the same as seen from any spacetime point: it is completely homogeneous. It follows that all spacetime points have exactly the same geome-

---

[2] We employ the convention that time-like distances are positive.

trical properties—again, it seems impossible to ground their individuality in their geometrical features. PII therefore seems to tell us that either there is only one point, or the individuality of the special relativistic spacetime points is of a haecceistic, geometry-transcending nature.

The transition to general relativity does not alleviate this dilemma. First, the famous "hole argument" strengthens the idea that the introduction of haecceities is incompatible with accepted physical thinking about spacetime. Indeed, if the existence of haecceities of spacetime points were to be accepted, this would in the context of the hole argument lead to a complete indeterminateness of the physical description: given one description, infinitely many other empirically fully equivalent descriptions can be produced that differ only by the way they are positioned in the spacetime manifold. These descriptions differ by virtue of the haecceities of the spacetime points that are involved, but are exactly the same as far as their geometrical properties and empirical predictions are concerned. The hole argument can in fact be summarized as the observation that if this numerical diversity of descriptions (and therefore also of solutions of the dynamical equations) is taken seriously, a radical indeterminism of physical theory has to be accepted: there are in this case infinitely many empirically equivalent solutions of the dynamical equations, given any initial condition.

From a physical point of view this miraculous multiplication of possibilities is just an unnecessary complication introduced by adding theoretical surplus structure. Haecceities are superfluous from an empirical point of view and constitute a paradigm case of objectionable metaphysics. But their rejection has the consequence that we are compelled to look for another basis of the individuality of spacetime points, presumably by using PII.

However, now our earlier problem repeats itself: also general relativity allows spacetimes with a high degree of symmetry. Minkowski spacetime, the limiting case of special relativity that is contained in general relativity, is only one example. The standard cosmological models, the so-called Friedmann-Lemaître-Robertson-Walker (FLRW) spacetimes, furnish further instances. These spacetimes allow the introduction of a cosmic time scale, at each instant of which three-dimensional space is completely homogeneous. Repetition of the earlier argument based on PII then would lead to the conclusion that at each instant of cosmic time there is only *one* spatial point (Wüthrich 2009). Evidently, this conclusion conflicts with the way general relativity itself describes these cosmological models.

3. Weak discernibility

As Hawley (2009) points out, defenders of Leibniz's principle can respond to such putative counterexamples to PII in a variety of ways. First, they may query whether the described situations are really possible at all—this does not appear to be a promising way out in our above examples, which are all possible according to the relevant physical theories. Second, defenders of PII can dispute that these situations are best described in terms of distinct but indiscernible individuals. This can take two forms: either it may be argued that a correct analysis of discernibility will reveal that the objects in question are discernible after all, or it may be claimed that it was a mistake to assume that there were distinct objects to start with—that there is actually only one undivided whole.

The latter option should certainly be taken seriously. It can be argued that the quantum case (the case of "identical quantum particles") calls for exactly this response: there are no individual fermions and bosons but there is rather only one undivided quantum field (Dieks and Versteegh 2008; Dieks 2010; Dieks and Lubberdink 2011). Later we shall say a bit more about the justification for taking this route. However, in the geometrical cases, and also in the cases of Black's spheres and Kant's hands, there are good reasons for going another way. As we shall argue, here we are entitled to think that we are in fact dealing with separate objects and that a more refined analysis of discernibility can show this to be justifiable by PII.

In order to explain this more refined analysis we follow Saunders (2003; 2006), who takes his cue from Quine (1981), in noting that in cases like the ones mentioned above *irreflexive* relations are instantiated: i.e., relations that entities cannot bear to themselves. Thus, Black's spheres are at a non-zero distance from each other—but a sphere cannot be at a non-zero distance from itself;[3] Kant's hands are each other's mirror images—but a hand cannot be its own mirror image; and the points in the Euclidean plane have distances with respect to each other that they cannot possess with respect to themselves. Similarly, in each of the spaces-at-an-instant in the FLRW cosmological models general relativity defines a spatial distance function such that points do not have non-zero distances with respect to themselves (at least as long as one restricts oneself to spatial regions that are not too large—since these spaces may be closed, the situation in the

---

[3] We are here taking it for granted that the distance relations satisfy Euclidean geometry. If, instead, the distances satisfied the geometry of a three-sphere, or another closed space, the argument given below for the numerical diversity of the spheres could fail.

large may be analogous to that of points on a circle). More generally, in arbitrary relativistic spacetimes irreflexive relations between points can be constructed on the basis of the four-dimensional distance function that relativity theory defines on such spaces (see also Muller 2011). In the case of points with space-like or time-like separation this is obvious: in this case there is a positive or negative distance between the points that cannot exist between any point and itself (at least not within regions that are not too large). The *prima facie* more complicated case of points with light-like separation (i.e. with four-distance zero) can be accommodated by observing that any two distinct points that are light-like separated can be connected via the combination of a non-zero time-like and a non-zero space-like interval,[4] something which cannot hold for any point with respect to itself.

The irreflexivity of these relations is the key to proving that (a generalized version of) PII is satisfied after all: if an entity stands in a relation that it cannot have to itself, there must be at least two entities.

To see in detail how this works, let us formalize the argument. PII can be formulated as follows, with = denoting identity:

$$s = t \equiv \forall P(P(s) \leftrightarrow P(t)). \qquad (1)$$

The universal quantifier here ranges over all *physical* predicates P (not haecceities!). The right-hand side of the equation stipulates that s and t can replace each other, *salva veritate*, in any P.

There can now be various kinds of discernibility (Saunders 2006). Two objects are *absolutely discernible* if there is a one-place predicate that applies to only one of them; *relatively discernible* if there is a two-place predicate that applies to them in only one order; and *weakly discernible* if an *irreflexive* two-place predicate relates them. The latter possibility is relevant to our examples. If there is an irreflexive but symmetric two-place predicate $P(.,.)$ that is satisfied by s and t, the definition (1) requires that if s and t are to be identical, we must have:

$$\forall x(P(s,x) \leftrightarrow P(t,x)). \qquad (2)$$

But this is false: in any valuation in which $P(s,t)$ is true, $P(t,t)$ cannot be satisfied since P is irreflexive. It follows that PII is satisfied by any two non-identical objects that stand in an irreflexive physical relation.

---

[4] Consider two points P and Q, such that the four-distance $\sigma(P, Q) = 0$. Then there exists at least one point R such that $\sigma(P, R) > 0$ and $\sigma(R, Q) < 0$.

In our cases with irreflexive relations PII is therefore sufficient to ground the numerical diversity of the objects after all. It should be noted, however, that although weak discernibility is thus able to lay a non-haecceistic fundament under numerical diversity, this does not endow the objects with identifiability in the usual sense. Indeed, in the situations we have discussed it remains impossible to pick out or define any single object. Because of the symmetry any property or relation that can be attributed to one object can equally be attributed to any other and we can therefore not identify any specific object. It is impossible, for example, to pin down any particular point in the Euclidean plane on the basis of the properties of the plane and its points, even if we include all relational properties (of course, we are speaking here of the plane *tout court*, without adding by hand a preferred point that could function as an origin).

This lack of identifiability may raise doubts on the meaningfulness of the claim that there are more than one separate objects after all (cf. Keränen 2001). However, in order that the number of elements in a domain is a well-defined quantity it is sufficient that a function *exists* that maps the domain one-to-one onto a set of labels, e.g. the set $\{1, 2, ..., n\}$; it is not required that we can actually *construct* such a labeling. In the examples we have been considering it was actually given in the description of the cases (two spheres, two hands, many points) that such mappings exist, and the question to be answered was simply whether this mapping could be given a basis via PII. As we have seen, this can indeed be done, with the help of irreflexive relations.

4. *Scientifically respectable objects and relations*

It is important, of course, that the irreflexive relations we consider here are scientifically respectable. If we started by just stipulating that a certain domain is labeled by a set of natural numbers, without providing a physical correlate to this labeling, this would be empty from a physical point of view. Labels can always be posited abstractly, and there are always irreflexive relations between them; for example, the irreflexive relation of "being unequal to each other" between the natural numbers. Proceeding along this road we would do nothing to make it acceptable that the domain in question really splits up and consists of different physical objects. The numbers would in this case function as haecceities in disguise, without scientific merit.

The justification for "splitting up a domain" is certainly not *a priori* evident in all cases in the context of discussions about whether or not PII applies. As we have seen at the beginning of the previous section, one possible stance in these discussions is to argue that there is no multiplicity at all: that there is only *one* undivided physical system. No questions about the individuation of component systems by means of PII have to be answered if there is only one system, and a parsimonious ontological picture results. If there is no convincing reason to think of the domain as consisting of several separate entities in the first place, this kind of holism surely recommends itself (Hawley 2006). As stressed before, what we need in order to think of numerical diversity is an argument that the objects and the relations between them are scientifically respectable and latch on to the structure of the domain as described by physics.

A note is in order here: it may be the case that there is nothing else to physically characterize the objects in the domain than the *relations* they stand in. This possibility has been frequently discussed in the recent literature and has led to the idea of structuralism, according to which objects are viewed as nodes in a relational network (this is compatible with the idea that the relations in their turn can only exist if they connect actual relata (Esfeld and Lam 2010)). So if we are going to verify scientific respectability we need not necessarily assume that there is a division of labor between "objecthood providers" that must be checked first, and relations that only become relevant later. A relational structure may constitute our only access to the existence of entities—in fact, this is the situation we shall encounter in the case of spacetime points.

5. *Spheres, hands, euros and other objects*

Looking back at the examples of Black's spheres and Kant's hands, we notice that the objects there (spheres and hands, respectively) were characterized independently of the irreflexive relations that hold between them. The spheres were described as being of a certain chemical composition (pure iron, says Black) and of course as possessing a definite geometrical shape. We know very well what such spheres are like: we have seen many objects like them and on this basis we can be sure that they are *bona fide* physical objects. Similarly, we are familiar with hands and cannot doubt their status as physical entities. The difficulty of these cases is not in deciding whether there are objects at all, but rather in finding an empirical basis for their numerical diversity: since the physical characteristics are completely the same for both objects, PII seems to suggest that there is actually only one of them. The information that an irreflexive relation is instantiated (being at a two-miles distance from each other and being each other's mirror images, respectively) now helps out: application of PII demonstrates that there must be two objects—although the perfect symmetry of the situation makes them only weakly discernible.

The uncontroversial physical nature of the irreflexive relations in question is highly relevant here. For when we allow relations that refer to "fantasy predicates," not sanctioned by physics, there is no limit to the number of entities that may result. One might for instance imagine that in every perfect iron sphere a "mork" and a "gork" copy of this sphere coincide, and that since these two qualities exclude each other (the irreflexive "mork is not gork" relation) we have actually two spheres instead of one. This example is simplistic and far-fetched, but the point still stands that we should make sure that the relations we are considering are *bona fide* from a scientific point of view if we wish to draw conclusions about the number of scientifically respectable objects. If the relations are our only access to the objects, it becomes even more pressing to verify their scientific status.

A well-known more realistic example of irreflexive relations without there even being any actual objects at all is the case of money in a bank account (not coins in a piggy bank, but transferable money in a real bank account). Imagine an account with five Euros. It is easy enough to *speak* about this money as a collection of five entities, and a mapping of the set {1,2,3,4,5} to the account can be defined in an abstract way (e.g., draw a circle representing the total amount and divide it into 5 equal parts; or think of the mapping as given by the order in which amounts of one Euro were transferred to the account), but this does not prove anything about the actual presence of five individual entities in the account. On the contrary, the case of more than one money units in a bank account is the standard example of absence of individuality; it is a case in which only the account itself, with the total amount of money in it, can be treated as possessing individuality (Schrödinger 1952; 1998; Teller 1998). Although we are accustomed to using relations and things *talk* in this case (e.g., "the last Euro that has come in is more important than the other ones"), this talk does not represent the actual physical situation (it may represent aspects of how that situation came about; but that is not our concern here).

A (more controversial!) example from present-day science comes from quantum mechanics. A notorious interpretational issue in this theory is the status of so-called "identical particles": are these "particles" individual entities or should they rather be described in a holistic way (in terms of the state of one field)? The theory suggests that there are distinct entities because it works with indices 1, 2, 3, ..., n that label the one-particle Hilbert spaces (the quantum mechanical state spaces) that can be used to construct the total state space of the system. However, it is a general principle of quantum mechanics that the state defined in this total space must be completely symmetrical (the case of "bosons") or anti-symmetrical ("fermions") in these indices. This symmetrization postulate implies that if the total state is restricted to any of the one particle state spaces the result is exactly the same for all indices.[5] It follows that if these indices are to represent particles, any property or relation that may be attributed to any one of them is attributable to each of the others as well.

One response is to take this as a signal that there are no different particles at all: although there is talk about the indices in the formalism in terms of particles, this should be understood in the same way as talk about the different Euros in a bank account. From a fundamental point of view it is better, according to this line of reasoning, to renounce talk that suggests the existence of individual particles and to reconceptualize the situation in terms of the excited states of a field (analogous to thinking of the Euros in an account as one sum of money). This response leads into the direction of quantum field theory.

However, the situation is also reminiscent of Black's spheres and Kant's hands. As we have seen there, symmetry is not decisive for proving the absence of Leibniz-style individuality: we may be facing a case of *weak* discernibility. Perhaps there are irreflexive physical relations between the fermions and bosons that guarantee their (weak) individuality in the same way as they did for Black's spheres and Kant's hands (Muller and Seevinck 2009, Saunders 2006). Now, we have already seen that in the total state indices 1, 2, 3, ..., n occur; and it is certainly possible to define irreflexive relations between them (see Dieks and Versteegh 2008, Dieks 2010, Dieks and Lubberdink 2011, Muller and Seevinck 2009, Saunders 2006 for details). But does it follow that these indices indeed label weakly discernible physical entities?

---

[5] The restricted state is found by taking the "partial trace".

6. To be or not to be

Obviously, we know that spheres and hands are trustworthy physical objects: we possess direct experience of objects of this kind. But let us reflect for a moment about the background of our certainty in these cases. We are familiar with collections of spheres and hands in *asymmetrical* situations, where it is possible to uniquely distinguish and name them. Typically, we think of two hands or two spheres as placed differently with respect to ourselves as observers: one being at our left and one at our right, for example. Such asymmetries make the objects *absolutely* discernible: they obtain their own distinguishing features that make them identifiable. We can grab and name them in such circumstances, and direct others to them by giving identifying descriptions.

The possibility of these asymmetrical configurations thus gives us confidence about the nature of spheres and hands even if they are placed in a completely symmetrical situation. They are clearly entities of which several copies can exist next to each other. The symmetrical configurations in which they are only weakly discernible are limiting cases of more typical asymmetrical situations in which a point of reference, an observer, or something similar is given, and in which they are absolutely discernible.

Likewise, the typical context in which classical particles occur is that of asymmetrical situations, in which the network of mutual distances suffices to characterize each individual particle in an unambiguous way. Changing the mutual distances so that the configuration becomes more symmetrical will evidently do nothing to the nature of the objects: as long as the situation is only slightly asymmetrical they will remain absolutely discernible physical entities, whereas in the limiting case of complete symmetry they still are the kind of entities that are *candidates* for absolute discernibility. There is no indication in physical theory or anywhere else that there might be ontological changes in the objects just because of approaching a fully symmetrical Black's spheres-type configuration.

The (standard, absolute) discernibility in asymmetrical situations, plus the possibility of a limiting procedure, thus provides us with a test for physical relevance and physical objecthood in symmetrical situations. We are justified in assuming the existence of actual entities (candidate individuals) if the breaking of the symmetry is physically possible, does not involve any change in the type of physical properties assigned, and results in a situation with absolutely distinguishable objects (this strategy to some extent resembles the one followed by Adams, who proposes to compare

Black's spheres with spheres of which one has a very slight chemical impurity (Adams 1979, 17)).

These cases are to be contrasted with the case of the Euros in a bank account. If we start with actual Euro coins, there is no limiting procedure by means of which we can gradually approach Euros as units of transferable money. There is consequently no argument here that bank account Euros and Euro coins are ontologically similar. On the contrary, according to our best available way of describing what is going on in bank accounts, there is only a total amount of money in them, not composed of individual Euros.

To investigate into which category "identical quantum particles" fall, the transferable Euros or the Black's spheres, we can try and copy the strategy followed in the classical particle case, namely breaking the symmetry and seeing whether absolutely discernible entities result. But here we run into a difficulty of principle: quantum mechanics *forbids* "identical particle" systems that are not in a fully (anti-)symmetric state—it is a matter of quantum mechanical law that "fermions" and "bosons" can only have exactly the same states and relations. (More accurately: the *indices* occurring in the formalism must always be completely interchangeable in formulas that express observable quantities.) This is significantly different from the symmetrical classical cases, in which the symmetry was contingent and the theory allowed evolutions from symmetrical to asymmetrical configurations. In quantum mechanics the mutual relations between fermions cannot serve to distinguish individual component systems *as a matter of principle*, and our earlier test fails.

It might be replied that this by itself does not yet prove that there are no individual identical quantum particles—indeed, we could also imagine a hypothetical *classical* world in which a law stipulates that perfect spheres can only occur in completely symmetric configurations. But in such a world we could still have good reasons to think in terms of individual spheres: our theories could allow for an external object serving as a point of reference that makes the spheres discernible (e.g., an observer who stands in different relations to the various spheres). If no argument of this kind were to be possible at all, the existence of individual spheres would surely become moot. However, this is precisely the situation that obtains in quantum mechanics. It follows from the quantum formalism, as a matter of law-like principle, that "identical quantum particles" have exactly the same relations with respect to any external vantage point that may be introduced (Dieks and Versteegh 2008; Dieks 2010).

More can and should certainly be said about the quantum case (cf. Dieks and Lubberdink 2011)—the subject remains controversial. But the above should suffice to show that it is not at all evident that "identical quantum particles" should be conceived of as numerically diverse objects.

7. The status of spacetime points

Where does this lead us with respect to the question of whether spacetime points are individuals—and if so, in what way their individuality can be grounded? In section 2 we have encountered the dilemma that the points in the Euclidean plane have all their monadic and relational properties in common, as a consequence of the perfect homogeneity of the plane. If we wish to avoid haecceities and try to ground the individuality of geometrical points in their geometrical properties, this appears to lead to the conclusion that there cannot be more than one point in the plane. The situation does not seem to improve when we go from mathematics to physics: in Minkowski spacetime, but also in symmetrical solutions of the field equations of general relativity (e.g., the FLRW models), spacetime points share again their properties and the conclusion that on a non-haecceistic account this entails that the universe consists of one lonely point has indeed been drawn in the literature (Wüthrich 2009—the author of this article intends this as a *reductio* of a non-haecceistic structuralist position).

As we have explained in section 3, the notion of *weak discernibility* can come to the rescue here. Both in the case of mathematical geometries and in the case of physical (spacetime) geometries irreflexive relations exist between the points. If these relations are admitted to the domain of relations used in the application of Leibniz's principle PII, logic dictates that a numerical diversity of weakly indiscernible objects exists, exactly as needed.

However, we should not be too quick. As we have seen illustrated in sections 4, 5 and 6, we need some guarantee that the thus defined weakly discernible objects, and the relations between them, are scientifically respectable: we should restrict the domains of quantification in PII so as to avoid artificial predicates and properties.

When we look again at the mathematical example of Euclidean geometry (in a modern axiomatization), the first thing to observe is that the geometrical points are not assumed to possess any properties except for what is fixed by the relational structure to which they belong. It is possible to axiomatize the full geometry of the plane by describing it as a manifold of elements on which a non-negative distance function is defined. So what we have here is the structuralist possibility already alluded to at the end of section 4, in which our only access to objecthood is provided by the relations between the objects. In the present case these relations are mutual distances—and since distances greater than zero constitute irreflexive relations we recover the usual structure of the plane with its infinity of points as weakly discernible objects.

Now, as long as we are discussing the plane as a purely mathematical structure, the question of whether the distance relations are scientifically significant or not and of whether it would not be better to describe the plane as one undivided whole, does not arise: within mathematics the relational structure is just *posited* to exist (either as being defined by us or as abstractly existing in a mathematical sense) and studied as such.

The interesting questions pertain to *physical* respectability—even if a mathematical structure with numerical diversity is perfectly alright in itself, it can still be asked whether its application to physical reality is justified. So consider a physical counterpart to the Euclidean plane, namely the completely homogeneous three-dimensional space that occurs in Newtonian physics. Here also all points have exactly the same status, and can be understood as specified by the structure of distance relations in which they are the relata. This leads again to an (uncountable) infinity of weakly discernible points. By contrast to the purely mathematical case, however, we may now raise the question whether the network of distances, and the points in it, are in fact physically significant.

At first this may seem an absurd question: what could be more physically significant than distances? However, we were considering *empty* Newtonian space, without material contents. That means that there are no measuring rods or physical processes that "feel" the distances. The introduction of such material things would generally destroy the homogeneity that we have assumed, and would lead to a completely different situation. Now, if Newtonian physics were nothing but the study of empty Newtonian space, the numerical diversity of the points in this space would fulfill no physical role. In this case we would have no empirical warrant that the relational structure and the points in it correspond to anything physical, and it would be fully justified to think of space as one undivided entity. In other words, if only a completely empty homogeneous space were physically possible, its internal structure would have no obvious physical meaning. It could be stipulated to exist, of course, but its status would be mathematical or metaphysical rather than physical.

But this is evidently not at all the predicament we are actually in. Real-life applications of Newtonian physics deal with situations that are not even nearly symmetrical, because they assume many objects with unequal properties in asymmetrical configurations. In such situations it generally becomes possible to discern the spatial points *absolutely*, with the help of distances and angles with respect to the discernible objects that occupy points in the space. This is analogous to what happens when we add an origin and coordinate axes to the Euclidean plane. The completely symmetrical homogeneous case thus appears as an idealization, a limiting case that results from further and further abstracting from actual asymmetries. If we look at all physically possible particle configurations in Newtonian space there can be no doubt about the physical significance of the (absolutely discernible!) points: in most configurations there will be actually physically instantiated, distinguishing relations.

The physical significance of the relations and relata in the fully symmetrical cases can thus be motivated by a comparison with other possible cases, in which there are *asymmetries*. Newtonian theory (like other spacetime theories) implicitly assumes that the basic ontological nature of spacetime is the same regardless of the exact physical situation that is being described. This justifies the comparison of spacetime on its own, being empty, with spacetime filled by complicated asymmetrical physical particle and field distributions. Since it is clearly physically meaningful to consider spacetime points as individuals in the latter case, we are justified in using this picture also in the vacuum case. In other words, there is a *modal* aspect involved in the reasoning here, since we compare different physical situations that are possible according to the same physical theory (so that the modality that is relevant here is *physical*, determined by physical theory).

Completely similar observations can be made for special and general relativity. If special relativity were exclusively a theory about empty Minkowski spacetime, the physical significance of spacetime points in the theory would be doubtful. And if general relativity were solely about completely symmetrical FLRW universes the same would be true here. But in fact empty Minkowski spacetime and the completely homogeneous FLRW cosmological models are degenerate cases, that can be connected to more realistic, asymmetric spacetimes via a limiting process (albeit an abstract one, that goes from one physically possible situation to another). This is just like what we saw in the cases of Black's spheres and Kant's hands: the physical significance comes from a comparison with asymmetric situations.

8. Conclusion

In completely symmetrical situations Leibniz's principle PII seems *prima facie* unable to ground the individuality and numerical diversity of objects. The notion of weak discernibility, using irreflexive relations, saves the day: although it cannot yield individuality in the ordinary sense of identifiability, it is able to ground the numerical diversity of objects. In particular, this seems to offer a way out of the problem of how to individuate points in space or spacetime.

What we have argued in this article is that we need a criterion to ensure that the weakly discernible objects that thus result are scientifically respectable. In order to verify this respectability we have to make sure that the irreflexive relations upon which the weak discernibility hinges are scientifically relevant. In particular, in the space and spacetime cases we need an argument that the mutual distances between space or spacetime points are physically significant. As it turns out, this physical significance depends on the possibility, according to physical theory, of situations in which the symmetry is *broken*: we can only have empirical access to single objects (here: points) via relations that absolutely distinguish different objects. It is this absolute distinguishability that assures us that we are dealing with objects at all.

If this is correct, the relevance of the irreflexive relations used to establish *weak* discernibility depends on their ability to ground *absolute* discernibility in asymmetrical situations. Consequently, weak discernibility only helps out if it can be regarded as a degenerate case of absolute discernibility.

REFERENCES

Adams, R. M. (1979). Primitive Thisness and Primitive Identity. *Journal of Philosophy*, 76, 5-26.

Black, M. (1952). The Identity of Indiscernibles. *Mind*, 61, 153-164.

Dieks, D. (2010). Are 'Identical Quantum Particles' Weakly Discernible Objects? *EPSA, Philosophical Issues in the Sciences* (Chapter 3), Vol. 2, eds. M. Suárez, M. Dorato and M. Rédei. Dordrecht: Springer.

Dieks, D., and Lubberdink, A. (2011). How Classical Particles Emerge From the Quantum World. *Foundations of Physics*, 41, 1051-1064.

Dieks, D., and Versteegh, M. A. (2008). Identical Quantum Particles and Weak Discernibility. *Foundations of Physics*, 38, 923-934.

Esfeld, M., and Lam, V. (2010). Moderate Structural Realism about Space-Time. *Synthese*, 160, 27-46; Ontic Structural Realism as a Metaphysics of Objects, *Scientific Structuralism* (Chapter 8), eds. A. Bokulich and P. Bokulich, Dordrecht: Springer.

French, S., and Krause, D. (2006). *Identity in Physics: A Historical, Philosophical, and Formal Analysis.* Oxford: Oxford University Press.

Hawley, K. (2006). Weak Discernibility. *Analysis*, 66, 300-303.

Hawley, K. (2009). Identity and Indiscernibility. *Mind*, 118, 101-119.

Keränen, J. (2001). The Identity Problem for Realist Structuralism. *Philosophia Mathematica*, 9, 308-330.

Muller, F. A. (2011). How to Defeat Wüthrich's Abysmal Embarrassment Argument against Space-Time Structuralism. *Philosophy of Science*, 78, 1046-1057.

Muller, F. A., and Seevinck, M. (2009). Discerning Elementary Particles. *Philosophy of Science*, 76, 179-200.

Quine, W. V. (1981). Grades of Discriminability. *Journal of Philosophy*, 73, 113-116. Reprinted in *Theories and Things*, W. V. Quine, Cambridge, MA: Harvard University Press.

Saunders, S. (2003). Physics and Leibniz's Principles. *Symmetries in Physics: Philosophical Reflections*, eds. K. Brading and E. Castellani, Cambridge: Cambridge University Press.

Saunders, S. (2006). Are Quantum Particles Objects? *Analysis*, 66, 52-63.

Schrödinger, E. (1952, 1998). *Science and Humanism.* Cambridge: Cambridge University Press. Partly reprinted as: What is an Elementary Particle? *Interpreting Bodies: Classical and Quantum Objects in Modern Physics*, ed. E. Castellani. Princeton: Princeton University Press, 197-210.

Teller, P. (1998). Quantum Mechanics and Haecceities. *Interpreting Bodies: Classical and Quantum Objects in Modern Physics*, ed. E. Castellani. Princeton: Princeton University Press, 114-141.

Wüthrich, C. (2009). Challenging the Spacetime Structuralist. *Philosophy of Science*, 76, 1039-1051.

A STRUCTURAL AND FOUNDATIONAL ANALYSIS OF EUCLID'S PLANE GEOMETRY: THE CASE STUDY OF CONTINUITY

PIERLUIGI GRAZIANI
University of Urbino
Department of Basic Sciences and Foundations
pierluigi.graziani@uniurb.it

ABSTRACT. In this paper I offer a structural[1] and foundational analysis of Euclid's Plane Geometry. These objectives require the analysis of some characteristics of Euclid's plane geometry and of the logic of his argumentations. In particular, I will focus my analysis on the continuity problem as a case study. So, the aims of this paper are: (a) to show that it is necessary to analyse the status of the continuity postulate in the *Elements*; (b) to contribute to an understanding of how the *Elements* contain particular cases of the continuity postulate that are certainly used by Euclid, even though they are not explicitly stated; (c) to contribute to an understanding of how these particular cases really play a fundamental role, due to the centrality of diagrammatic configurations in the *Elements*; (d) to contribute to an understanding of how the study of these issues can also be important for developing modern synthetic-constructive axiomatizations of Euclid's geometry; (e) to present a Sequent Calculus for Euclid's Plane Geometry.

1. A famous example

We can start by considering a (famous) example in which the continuity problem is of great relevance.

Problem I.1 of Euclid's *Elements* requires us to 'construct' an equilateral triangle on a 'given finite straight line', or segment, in modern parlance. Let's see the Euclidean solution:

---

[1] In proof-theoretical sense.

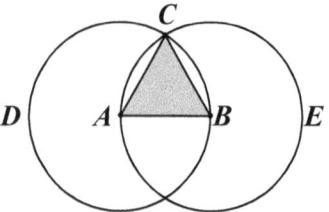

Protasis: To construct an equilateral triangle on a given finite straight line.

Ekthesis: Let AB be the given finite straight line.

Diorismos: It is required to construct an equilateral triangle on the straight line AB.

Kataskeue: Describe the circle BCD with center A and radius AB. Again describe the circle ACE with center B and radius BA. Join the straight lines CA and CB from point C at which the circles cut one another to points A and B. (Post. 3; Post.1).[2]

Apodeixis: Since point *A* is the center of circle *CDB*, therefore *AC* equals *AB*. Again, since point *B* is the center of circle *CAE*, therefore *BC* equals *BA*. (I. Def. 15). But AC was proved equal to AB, therefore each of the straight lines AC and BC equals AB. And things which equal a same thing also equal one another, therefore AC also equals BC. (C.N. 1).

Sumperasma: Therefore the three straight lines AC, AB, and BC equal one another.[3]

So, to achieve his result, Euclid takes this segment to be AB, then d scribes two circles with centres in the two extremities A and B of this segment, respectively, and takes for granted that these circles intersect each other in a point C. This is not licensed by his postulates. Hence, either his argument is flawed, or it is warranted on other grounds.

According to a classical view 'the Principle of Continuity' provides an appropriate ground for this argument, insofar as it ensures 'the actual existence of points of intersection' of lines.

Michael Friedman[4] has remarked, however, that in the *Elements* "the notion of 'continuity' [...] is not logically analyzed" and thus there is no

---

[2] Post. = Postulate; Def. = Definition; C.N. = Common Notion; Prop. = Proposition.

[3] For a study of the different parts in which Greek demonstrations and solutions are conventionally divided you can see Acerbi 2011; Sidoli and Saito 2012.

room for a "valid syllogistic inference of the form: C1 is continuous, C2 is continuous, then C exists", where C1 and C2 are the two circles involved in this proposition.

But we can see circles cutting one another in the diagrammatic configuration!

This example is very famous because it has played a central role in the construction of the classical interpretation of continuity in Euclid. Here are some quotes.

Thomas Heath:[5]

> It is a commonplace that Euclid has no right to assume, without premising some postulate, that the two circles *will meet* in a point C. To supply what is wanted we must invoke the Principle of Continuity.

Bertrand Russell:[6]

> [...] Countless errors are involved in his first eight propositions. That is to say, not only is it doubtful whether his axioms are true, which is a comparatively trivial matter, but it is certain that his propositions do not follow from the axioms which he enunciates. A vastly greater number of axioms, which Euclid unconsciously employs, are required for the proof of his propositions. [...] as Euclid fails entirely to prove his point in the very first proposition. As he is certainly not an easy author, and is terribly long-winded, he has no longer any but an historical interest. Under these circumstances, it is nothing less than a scandal that he should still be taught to boys in England. A book should have either intelligibility or correctness; to combine the two is impossible, but to lack both is to be unworthy of such a place as Euclid has occupied in education.

L. M. Blumenthal:[7]

> [...] his (Euclid's) foundation was simply insufficient to support the lofty edifice he sought to erect on it. He was able to prove many of his theorems because he used arguments that cannot be justified by his postulates. This occurs in the very first proposi-

---

[4] Friedman 1985, 60.
[5] Heath 1908, 242.
[6] Russell 1917, 108.
[7] Blumenthal 1961, 3-4.

tion, which purports to show that on any given segment A, B an equilateral triangle can be constructed. His proof is invalid (incomplete), inasmuch as it makes use of a point C whose existence is not, nor cannot be, established as a consequence of the five postulates, since the circle with centers A and B and radius AB that he employed in his argument cannot be shown to intersect to the given point C. This difficulty is due to the lack of a postulate that would insure the continuity of lines or circles.

R. Hartshorne:[8]

As we read Euclid's Elements let us note how well he succeeds in his goal of proving all his propositions by pure logical reasoning from first principles. We will find at times that he relies on 'intuition', or something that is obvious from looking at a diagram, but which is not explicitly stated in the axioms. For example, in the construction of the equilateral triangle on a given line segment AB how does he know that the two circles actually meet at some point C? While the fifth postulate guarantees that two lines will meet under certain conditions, there is nothing in the definitions, postulates, or common notions that says that two circles will meet. Nor does Euclid offer any reason in his proof that the two circles will meet.

S. Feferman:[9]

Dedekind style continuity considerations only emerged in the critical re-examination of the Euclidean development in the 19th century when it was recognized that several of its proofs assume the existence of certain points that are 'supposed' to be there, but whose existence is not guaranteed by the Euclidean postulates. In fact, this occurs at the very outset in proposition I.1, which asserts that an equilateral triangle can be constructed on any line segment AB as base. This is proved by constructing two circles C1 and C2, with centers A and B, respectively, and radius equal to AB. The vertex of the equilateral triangle to be constructed is taken as either point of intersection of C1 and C2. But there is nothing in the Euclidean postulates which guarantees that such points exist. From the modern point of view, the circles may be 'gappy'. But that way of thinking is entirely foreign to Euclidean geometry, as is the general continuity axiom formulated in Hilbertian geometry below. Some modern developments of Euclidean geometry add

---

[8] Hartshorne 2000, 29-30.
[9] Feferman 2008, 8-9.

special continuity axioms ('line-line', 'line-circle', 'circle-circle') asserting the existence of points that 'ought to exist', and indeed evidently exist in our basic picture, but don't follow from the Euclidean postulates. (For details, see Greenberg (2007) or Hartshorne (1997).) One can imagine that Euclid himself might have acknowledged the logical necessity of adding such specific continuity postulates in one form or another if that were pointed out to him, but one cannot say that consideration of these questions is part of the basic picture.

Is the situation really like the one described above?

Objectively Euclid does not explicitly state a continuity postulate or postulates about the intersections of circle and circle (and circle and straight line), but he has not even developed a naïve discussion about continuity as someone tends to think. He, I will point out, has set out in the *Elements* the necessary conditions for the above mentioned intersections to exist.

Let us try to understand how. I will start by considering the general structure of Euclid's *Elements*.

2. Configurational interpretation vs. deductive interpretation

It is common knowledge that the *Elements* was not the first opus of its kind, but it was certainly the first to present much of the knowledge about elementary mathematics developed up to Euclid.[10]

The *Elements* comprise 13 books: I-VI elementary geometry; VII-IX number theory; X incommensurable quantities; XI-XIII solid geometry.

In Book I we can find many of the principles on which the opus is based: the *Definitions* (*horoi*), the *Postulates* (*aitemata*) and the *Common Notions* (*koinai ennoiai*). After declaring such principles, the opus proceeds directly to the statement and solution/proof of 48 propositions, which are divided into two kinds: those which describe a task (*problems*), and those which make an assertion (*theorems*).

The interpretation of the Postulates has been traditionally studied in the context of the *existential interpretations of geometric constructions*, i.e. *of the idea that Greek mathematicians would use geometric constructions to prove the existence of the constructed figures*. This interpretation, which

---

[10] Heath 1908.

has been usually associated to H. G. Zeuthen,[11] initially seemed to be logically and historically reliable, until A. Frajese, E. Stenius, W. Knorr and O. Harari (to mention just the fundamental ones)[12] showed its limits and provided a more adequate interpretation of the Euclidean geometrical construction and, therefore, of the Postulates.

When we analyze the meaning of the five Postulates we can see, at first, that the first three Postulates directly express the *possibility*, or a *required (postulated) ability, to execute specific constructions*: segments, extensions of finite straight lines, and circles. On such base, the Postulates were often said to correspond to the usage of ruler and compass; however, in Euclid there is no mention of such instruments, and a strictly instrumental description would require much more than what is stated by the Postulates which, to the contrary, in our opinion, express *idealized capabilities*. Euclid presents the Postulates as *minimal assertions* which can be enriched instrumentally by further propositions.[13]

Postulates IV and V also have, in Zeuthen's interpretation, a constructive nature: the former establishes the uniqueness of the extensions of finite straight lines required by Postulate II; the latter expresses the condition under which the intersection point of two straight lines can be built.

However, Zeuthen's interpretation of the Postulates does not seem to be coherent: for example there does not seem to be any reason why Postu-

---

[11] Zeuthen 1986.

[12] You can see references in bibliography. Specifically, Frajese's merit was to define a more coherent constructive interpretation of the Euclidean postulates; Stenius' was to give a subtle epistemological reading of procedures of geometrical constructions; Knorr's was to show how not all ancient geometrical constructions were driven by the need of existential proofs and, in addition, how Greek mathematicians were able to treat existential problems in many ways: postulates, tacit assumptions (such as geometrical continuity), existential theorems and also constructions; Harari's merit was to clarify, and diminish, the historical credibility and theoretical suitability of the standard existential interpretation (Zeuthen) of geometrical constructions, outlining with strength the sharp distinction between the Aristotelic concept of to be and the modern concept of existence, and between the Aristotelic and Euclidean conception of the geometrical objects; between the form of reasoning that Aristotle applies to geometrical issues and the proofs/solutions that Euclid gives in the *Elements*; and therefore between the modern concept of existence and the meaning of the Euclidean geometrical constructions.

[13] For example Proposition I,2 and Proposition I,3 enrich Postulates III by completing it and allowing the shifting of the segment which is necessary to give the Postulate an instrumental reading. For the exact analysis see Frajese 1950; and also Sidoli 2004.

late II should have such a completion by means of postulate IV and the others should not.

Frajese, in contrast, reduces the meaning of such Postulates and their constructive value to *the problem of the equality of figures.*

> Those geometrical figures which for Plato should be above all object of contemplation are instead for Euclid object of study and of geometrical considerations. But, in order for the figures themselves to become object of study, they first need to be compared to each other, i.e. a link between them needs to be built. In modern mathematics the concept of correspondence produces links, so the elements of a set are connected to each other by means of the structure that the set receives, i.e. by means of operations defined with such properties. For Euclid, figures can be connected by means of constructions and other means that one must be able to use: requirement which, according to us, would be made in the Postulates. So in Postulate I any pair of points would be connected by means of a segment of straight line; the extensions of finite straight lines (Post. II) would also allow us to reach (connect) the regions which are farther in the plain. A special link between the straight lines is their intersection: the conditions for this intersection to happen are stated in Postulate V. But, in those conditions, equality and inequality considerations come into play (comparisons which also connect the figures). Postulates III allows to recognize the equality (and also the inequality) between segments (the circle is a figure which allows for example to recognize that two segments are equal if they are radii of the same circle or if they can be reported to the radii).
>
> We shall see that in Proposition 2 and 3 of Book I, Euclid adds something to what is allowed by Postulates III, and explains how to shift the segments.
>
> For angles on the other hand one cannot generally recognize the equality by constructively executing the shift as it is possible with segments. Euclid will resort (though somewhat unwillingly and exceptionally) to a kind of mechanical transportation (a real mechanical movement) in Propositions 4 and 8 of Book I. But for at least one kind of angles (the straight ones) Euclid can postulate equality; this is what he does in Postulate IV.
>
> Between the straight angles, in other words, wherever they be located in the plane, some kind of remote connection is made: we could call it a radio connection, or wireless, as opposed to the wired connection i.e. with the straight lines of Postulates I and II, and with arcs of circles in Postulate III.[14]

---

[14] Frajese 1950, 302.

I believe there is good evidence to remain within the boundaries of Zeuthen's interpretations, at the same time accepting Frajese's interpretation as an integration. Frajese's interpretation is a *configurational interpretation* of Postulates and more in general of geometrical constructions: the mathematician works on configurations of objects by connecting their elements to each other with *interdependencies*, i.e. through the *construction of links highlighted by means of geometrical constructions.*

Euclidean Postulates have, therefore, a very refined synthetic-constructive dimension: they express *epistemic capacities* with which one can construct *geometrical links* between objects satisfying some condition, where the act of constructing is more similar to *producing evidence for the existence* of those geometrical links, than to generating them in strict sense.

Now, as was noted by Harari,[15] Euclid considers geometrical elements under two different perspectives: *the material* or *quantitative aspects* and *the positional* or *qualitative aspect*. In Book I of the *Elements*, Euclid gives two characterizations of geometrical entities: in Definition 1.1, for example, a point is defined as "that which has no parts" and in Definition 1.3 it is said that "the extremities of a line are points;" for lines we find the characterizations "breadthless lenght" and "extremities of a surface". Double characterizations can also be found for the other entities. So, these definitions can be classified (1) as definitions which determine quantitative aspects of geometrical objects (the former ones); (2) definitions which determine qualitative aspect of geometrical objects (the latter ones). The definitions in the former group distinguish the measurable from the non-measurable aspects and refer to the notion of divisibility as a way to determine the measurable aspects of the spatial object: a point is a non-measurable entity because it has no parts which can measure/quantify it; or a straight line is measurable with respect to its length, but not to its width. The definitions in the latter group characterize those aspects that are left outside by the other kind of definitions, i.e. the non-measurable aspects: points, for example, in the case of the line, are limits of the quantitative object. This is therefore the distinction between the material aspect (the divisible or measurable aspects) and the positional aspects (considered as a limit of the measurable figure) of geometrical figures. The priority of the material/quantitative dimension (characterization) over the positional/qualitative shows that Euclid (as opposed to the Pythagorean approach[16]) does not consider spatial positions as properties of objects, and that the notion of

---

[15] Harari 2003, 18-19.
[16] The Pythagorean definition of a point is "a unit having position".

geometrical space is not presupposed by geometrical objects. This means that the Euclidean space is not given, but *constructed* in sense of *made evident*: *it is a space where geometrical relations and geometrical objects are generated by building links*[17] *and limiting divisible magnitudes (materials).*

Geometrical constructions can therefore be interpreted as *epistemic procedures* (the acts of building links and of limiting divisible magnitudes) with which we can *construct* geometrical objects that satisfy certain conditions, where by "construct" we mean *to supply evidence to the existence* of objects of such kind instead of generating them ontologically. The Postulates of construction can be interpreted more precisely as stating *basic epistemic capacities, capacities to execute basic procedures.*[18]

What emerges above all from such investigation is the profound *configurational and relational nature* of the geometrical figures and mathematical procedures with which we prove theorems and solve problems. A geometrical figure is not seen in the perspective of its essential attributes (as in Aristotle), but in the perspective of its relationships with the other elements of the configurations it belongs to. Proofs and solutions must not be considered *only* as related to some steps of deduction from primitive elements to the sought conclusion but, *above all*, as related to the investigation of specific type of *geometrical configurations* where the search for proofs and solutions must be conceived as a study of the (*functional*) *interdependencies in the configuration between the geometrical elements which comprise it.*

In the light of such interpretation it is possible to claim, along with Harari, that geometrical constructions play a double role in the development of Euclidean propositions: (A) they serve as a measuring tool through which quantitative relations *are deduced*; (B) or they serve as instrument *to exhibit* qualitative relations, i.e. the order or the position of geometrical figures. The deductive employment of constructions does no more than *make explicit* the content that is already given in the *ekthesis*; in the second employment, to the contrary, constructions act as instruments capable of *developing* the content given in the *ekthesis*, placing the different geometrical entities in different special relationships.

---

[17] This perspective is coherent with the absence of a point-construction postulates, points don't construct links, they do not have parts and they are limits of the others geometrical objects. Points have to be assumed and can be obtained indirectly from fifth Postulates or from Problem I,1 etc.

[18] Stenius 1978. See also Mäenpää and von Plato 1990.

For instance, we can observe the proof of theorem I.5.[19] In its *apodeixis*, Euclid's proof leads from conjunction (1) "line AF is equal to line AG and line AB is equal to AC" to conjunction (2) "the angle ACF is equal to angle ABG and angle AFC is equal to angle AGB". These propositions rest on the construction steps preceding the proof, yet the role played by the construction steps in the tacit inferences that lead to these propositions is different. Proposition (1) is nothing but a reiteration of the content, which is given in the setting-out stage and in the construction step. In putting forward the relation of equality between line AF and AG, the construction step serves as a means of measurement; that is to say, it determines the quantity of line AG in comparison to the quantity of line AF. By contrast, proposition (2) does not rest solely on the content, which is given in the setting out, the construction steps and the first congruence theorem. Rather, in order to derive proposition (2), we are forced to give a new meaning to the ingredients of the figure. That is to say, the application of the first congruence theorem requires an act of visualization, where a configuration of lines is treated *as* a figure of a certain sort. For instance, prior to the proof, line AB is considered to be a side of a triangle, while line BF is regarded as a part of the additional line BD. The application of the first congruence theorem to this proof requires an act of visualization, where line AB and line BF are treated as side of a triangle. This act of visualization introduces the triangles AFC, thereby giving a new meaning to the other components of the configuration; line AC does not play the role of the side of triangle ABC and line FC turns into the base of a triangle. Similarly, the meaning of the spatial configuration's components undergoes further modification in the stage of *apodeixis*. In this stage, line BF is detached from the whole AF and is treated as a side of the triangle FBC. The introduction of triangle FBC modifies the meaning of the other components of the configuration. That is to say, line FC, which served as the base of triangle FAC in the first part of *apodeixis,* turns into one of the sides of triangle BFC. As a result, line BC, which served as the base of the triangle ABC, at the end of *apodeixis* turns into the base of triangle BFC.

3. The continuity postulate

Now, the profound *configurational and relational nature* of the geometrical figures and mathematical procedures with which Euclid proves theo-

---

[19] Harari 2003, 20-21.

rems and solves problems aids to understand the nature of the *Elements*. In this perspective we can also understand the role of continuity in Euclid's *Elements*.

My position is that the continuity is an unexpressed postulate, or, as Attilio Frajese said: a *sixth postulate*. In particular it is evident in Problem I.1. The fact that the two circles intersect each other in a point C, so that ABC is the searched triangle, is not proved. This gap is evident to many scholars, but only Attilio Frajese analyzed it in the context of links between books I and III of the *Elements*.

What are these links?

At first, we can say that in the third book we do not find any definition of straight line cutting the circle: Euclid gives a *negative definition* of secant line. In the second definition he writes: "A straight line is said to touch a circle which, meeting the circle and being produced, does not cut the circle." So the tangent is not separate from the circle, but it is connected to it; *it reaches it* but *does not cut* it. Then the tangent is a *not-secant* straight line. Both the secant straight lines and the tangent reach the circle, but only the first divides it in parts by means of a cut. Euclid gives a similar definition for the tangent to a circle (Def. III.3). The notion of secant line is improved since its previous formulation in the third book. The second proposition, for example, says: "If two points are taken at random on the circumference of a circle, then the straight line joining the points falls within the circle."

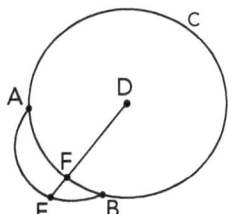

Proposition III.2 shows that the circle is a convex figure and also that finite straight lines cannot exist on the circumference: the straight line joining points A and B on the circumference does not fall on the circumference itself. It also shows that a straight line cannot have more than two common points with the circles' circumference: if a straight line had three common points, A, F, B, with the circumference, then DA=DF=DB would hold, and the external angle DFB would be greater than the internal angle not-adjacent FAD (FAD=FBD). Therefore DB > FD, but this is impossible.

The fact that the secant straight line and the intersected circumference have no more than two points in common does not imply that they have exactly two points in common: this idea is implicitly accepted. Euclid doesn't prove a sort of inverse proposition to the III.2: the theorem would begin with the hypothesis that a straight line passes through a point inside the circle; *the thesis would say that the straight line, suitably prolonged by the two parts, meets the circumference on two points.* But this thesis is a particular case of the continuity postulate.

Why didn't Euclid prove this evident theorem?

Attilio Frajese answered this question by showing that, although Euclid did not expressly enunciate the postulates concerning the intersections of circle and straight lines and of circle and circle, he nevertheless fixed the necessary conditions so that the intersections are there.

The proposition I.12 sheds light on the relation between circles and straight line.

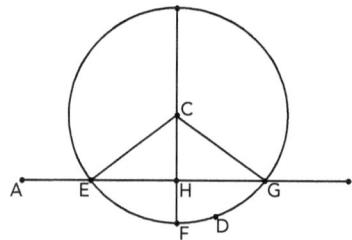

This proposition asks to draw a straight line perpendicular to a given infinite straight line from a given point not on it.

The construction I.12 requires that, by using the centre C and radius CD, we draw the circumference of a circle which cuts AB in two points: E and G. But what gives us the certainty that AB cuts the circle? CD, by joining two points in opposite places with respect to AB, cuts AB in K, and KC < CD. Therefore, K is inside the circle. The straight line AB can be prolonged as much as we like and therefore it contains two points outside the circle, and the straight lines joining these points and the point K cuts the circumference in E and G.

Therefore, an unexpressed postulate has a fundamental role in this proof. It is: "If a straight line passes through a point inside the circle, we can prolong this straight line from both ends and we can see that the straight line meets the circumference in two points." This unexpressed postulate is the inverse proposition of III.2, and although Euclid doesn't expressly enunciate it, the matter is correctly organized.

Euclid does not present a proof that circle and straight line cut each other under specific conditions. Euclid nevertheless organizes the matter so that the conditions to apply a particular case of continuity postulate hold.

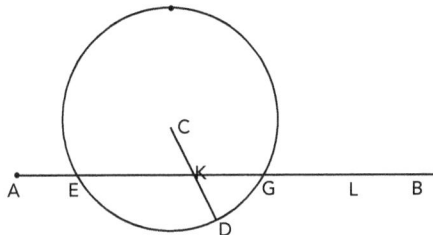

In I.12, as we have seen, Euclid organizes the matter accurately: the precaution to take a point D on the opposite place to C with respect to straight line AB, allows for AB to pass through a point inside to the circle: this point is K (the intersection between CD and AB).

The fact that CD cuts AB on point K follows from the fact that we have chosen D on the opposite place to C with respect to AB: K is inside the circle because CK < CD. But since the straight line AB is endless, we can find on it both a point K inside the circle, and another point distant from C more than the radius, for example L, and therefore outside the circle.

Euclid organizes the matter so that a circle passes through a point D taken on the opposite place to C with respect to straight line AB and then *a segment of AB has one point inside the circle and one outside the circle.* This way it is possible to satisfy a case of the continuity postulates: if a straight line pass through a point inside and a point outside to the circle, than straight line and circle cut each other.

Similarly to straight line and circle, Euclid proves in III.10 that a circle does not cut a circle at more than two points.

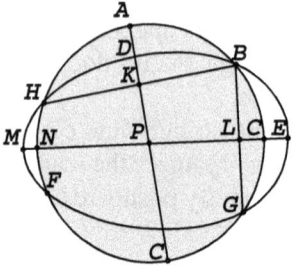

Now, it is exactly proposition I.22 which shows the condition such that two circles are secant.[20]

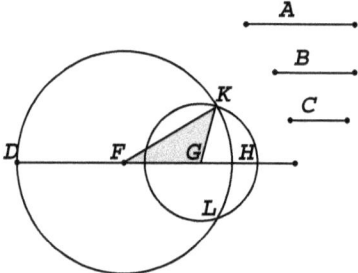

This proposition asks to construct a triangle out of three straight lines which equal three given straight lines: thus it is necessary that the sum of any two of the straight lines be greater than the remaining one [I.20].

In I.22 we have to admit that if specific conditions realize themselves (the distance of the centres is smaller than the sum of the radii and greater than their difference), then the two circumferences are secant. In other words: in I.22 we have to admit that if specific conditions realize themselves (the distance of the centres is smaller than the sum of the radii and greater than their difference), then it is possible to satisfy a case of the continuity postulate: *if both circumferences drawn pass through a point inside the other circle and a point outside the other circle, then the two circuferences are secant.*

---

[20] For an analysis of these propositions see Frajese 1950.

Therefore, evidently, a deep analogy exists between I.12 and I.22: both fix the conditions by which, respectively, a straight line cuts a circle and a circle cuts a circle.

4. Perspectives

Therefore, although Euclid has not expressly enunciated the continuity postulate, he was aware of the proposition; he doesn't leave the matter to human intuition, but organizes the matter so that some specific conditions can be given and then special cases of continuity postulate can be satisfied. But this organization and argumentation only make sense in a configurational interpretation of Euclid's *Elements*; for this reason I analysed how such interpretation *provides a fundamental ground to understand the Elements*.

To this point we can consider that a possible solution of the difficulty initially analyzed is to admit that Euclid's argument is configuration-based and that continuity provides a ground for it insofar as it is understood as a property of configuration.[21]

These ideas offer a new model to understand Euclid's mathematics and to offer a correct formal analysis of Euclid's argumentations and open new foundational research perspectives.

In particular I believe it is possible to give a positive answer to the following questions:
1) *is it possible to make the implicit assumptions and gaps that are present in Euclid's system explicit?*
2) *is it also possible to develop a new set of axioms, according to the Euclidean philosophy of mathematics and to modern standards of rigor, from which to deduce Euclid's synthetic geometry?*

To positively respond to these questions obviously means to account for continuity and other concepts which are implicit in Euclid's system such as those related to the method of superposition, the relative position of objects in the plane (a point which lies between other points; a ray which lies within a given angle), the concept of area, etc. Although in this paper we have focused our attention on the concept of continuity, the system that I will

---

[21] Marvelous papers on diagrams in Euclid's *Elements* are: Panza 2011; 2012.

present in the following paragraphs makes all the concepts involved in *Elements*' plane geometry explicit.[22]

4.1. Some preliminary ideas for a foundation of Euclid's geometry

As it can be observed in precedent chapters and by reading the *Elements*, Euclidean Postulates do not seem to be interpretable like *existential assertions in the ontological sense*, nor to play the role of what we could modernly call *existentially quantified propositions* or *existential axioms*.[23] If we consider the postulates of construction as a proposition of the kind $(\exists x)(Px)$, which states that at least one of the objects of given domain has property $P(x)$, we should read, say, the first Postulate as stating that between any two points there is a straight line (in the domain of straight lines) with the property of passing through those points. This does not seem to reflect the assertion made by the first Postulate, which, to the contrary, states something about *what we can do* in the sense of *epistemic capacities to execute procedures*. The observations given so far seem to support the idea of von Plato and Mäenpää[24] who, by *strengthening the relational interpretation of the Postulates,* achieve a *functional interpretation* of them. In such view, for instance, the first Postulate *highlights, by exploiting an ability, the existence of a functional relationship*, a function which takes two points as input and gives their connection through a finite straight line as output.

Under such interpretation, the first Postulate is much more complex than an axiom of existence because it does not explicitly talk about the existence of the set of all segments and it is not about the ontological attribution of existence but, to make a modern comparison, it reminds us of some kind of *Introduction Rule*[25] for such set, since it presents how its typical elements must be found. So, for some entities, there is no statement of existence in the ontological sense and the ability to study such domains is as-

---

[22] For an analysis of these other concepts see Hartshorne 2000, Graziani 2007.

[23] Proclus (1970, 77-82; 201, 5-9) introduced the theme of existential interpretation of geometrical problems by using Aristotelian terminology. In modern and contemporary times, the mathematicians generally consider the interpretation of problems as existential propositions and theorems as universal propositions. See von Plato and Mäenpää 1990, Mäenpää 1993.

[24] von Plato and Mäenpää 1990.

[25] In the sense of the *Introduction Rules* in the *Intuitionistic Type Theory*. See Martin-Löf 1984.

sumed by presenting how its elements can be constructed or produced, along with a statement of when two such elements are equal.[26]

The existential connotation of the Postulates and in general of construction is, therefore, more primitive than $(\exists x)(Px)$: it expresses something that is closer to the meaning of the judgment form $a:A$ in the *Intuitionistic Type Theory*[27], where such expression means that if *an object a of type A exists, an A exists in primitive sense.*

Therefore if, along the lines of von Plato and Mäenpää, we formalize the *Introduction Rule*[28] of segments as

$$[\text{lIntroduction}] \quad \frac{a:Point \quad b:Point}{l(a,b):Seg}$$

where a and b are points and their connection by means of a segment is expressed by $l(a,b)$, we observe that such rule expresses the first Euclidean Postulate more clearly than an *existential instantiation rule* could ever do.

For example, the following rule[29]

$$\frac{\exists x B(x)}{B(a)}$$

would reduce it to a mere exemplification (a means of representing a universal concept by introducing a specific corresponding instance); Von

---

[26] We can say that the fourth Postulate could be viewed as dealing with the equality of two canonical elements of *Straight Angle:Set*. The constructions of such equality are given in problems and theorems: Euclid, as we said, presents the Postulates as minimal assertions which can be enriched instrumentally by further propositions.

[27] Von Plato and Mäenpää use the language of the *Intuitionistic Type Theory* because the Predicate Logic (generally used) is not sufficient for a natural logical description of the configurational interpretation, of the construction postulates, and of the auxiliary constructions. They use Type Theory because it enriches Predicate Logic with a functional hierarchy that exactly captures, on the formal level, the Greek mathematical informal notion about these subjects, allowing at the same time a unified description of both the propositional dimension and the configurational dimension. For the same reasons I will use the language of the *Intuitionistic Type Theory* in the system that I present in the next section.

[28] The *Formation Rule* is: *Segment:Set*.

[29] Hintikka and Remes propose this interpretation in Hintikka and Remes 1976.

Plato and Mäenpää's rule also expresses the first postulate more clearly than

$$\frac{Point(a) \quad Point(b)}{Segment(l(a,b))}$$

where some individual properties are inferred from others so as to conceal any compositional and constructive dimension.

Von Plato and Mäenpää's rule, on the other hand, *establishes both a deductive connection between premises and conclusion, and functional dependencies between the constructions in the premises and the construction in the conclusion.*

This means that, when developing problems and theorems, we can have both deductive rigor and the ability to express deductions within a configuration made of functional links between the data, the conditions, the auxiliary constructions, and the thing sought.

Along those lines we are led to conclude that the usual interpretation of problems as existential propositions and theorems as universal propositions is not only historically but also conceptually inappropriate. A problem, in the Greek sense, projects the construction of certain sought objects in a specified relation to given objects. It had an infinitive grammar form requiring us to do something. Beside the practice of solving problems, the practice of proving theorems was also common among the Greek. A theorem is a requirement to demonstrate a property of a given object.

A problem in the Greek sense has three parts: the *given*, the *thing sought*, and the *condition*.[30] To solve a problem we have to construct the thing sought from the datum and we have to prove that the condition holds.

The intrinsic interest of the thing sought distinguishes a problem from a theorem, which has no thing sought. In order to prove a theorem, a mathematician will try to prove the condition for the datum.

It is important to note that the Greek would not distinguish the thing sought from the conditions: they used the term *zetoumenon* to refer to a combination of both.[31] This fusion creates ambiguity. The use of correct formal tools highlights this distinction.

---

[30] A problem may have several of each, or may in particular have not one of them.
[31] Mäenpää 1993 was the first to note this.

Therefore, the fundamental difference between what is contained in the assertion of a theorem and of a problem, as illustrated by Petri Mäenpää, can be formalized in the Theory of Types as

$$(y : B(x))C(x, y) : Problem \qquad (x : A)$$

where $x : A$ is the *datum*, $y : B(x)$ is the *thing sought*, and $C(x, y)$ are the *conditions*.[32]

To solve a problem we have to construct a $y : B(x)$ from $x : A$ and we have to prove that $C(x, y)$ holds.[33]

The existential proposition

$$(\exists y : B(x))C(x, y) : Problem \qquad (x : A)$$

can be used to express a problem with a condition. So problems are not identified with existential propositions.

Similarly, a theorem can be formalized as follows

$$C(x) \qquad (x : A)$$

which is an edge case for problems, since it lacks the thing sought. In order to prove a theorem, a mathematician will try to prove $C(x, y)$ for $x : A$.[34] A theorem can also be an existential proposition, for example when the condition is an existential proposition. We can find examples of existential theorems in Euclid's *Optic* (Theorems 37 and 38).

---

[32] As special cases: $(\exists y : B)C(y) : Prop$ (the datum is missing); $C(x) : Prop$ $(x : A)$ (the thing sought is missing); $B(x) : Set$ $(x : A)$ (the condition is missing).

[33] An example of a problem is the first Proposition of Euclid's *Elements*. Here the given object is a line segment, the sought object is a triangle, and the condition is that the triangle must be equilateral and construed on the line segment.

[34] An example of theorem is Proposition I, 32 of Euclid's *Elements*: it states that the angle sum of a triangle equals two right angles. Here the given object is a triangle and the condition is that the sum of its angles be equal to two right angles.

4.2. Foundation of Euclid's plane geometry

By considering previous analyses about continuity and nature of Euclidean constructions[35] I can now present a set of axioms and rules sufficient, by the modern standard, to provide a foundation of Euclid's Plane Geometry. This will lead even make explicit all those concepts (continuity *in primis*) that Euclid had as implicit in his system and that he relegated to an intuitive understanding.[36]

The system developed here has a great debt on the one hand with Jan von Plato's[37] research on constructive geometry and the other with Robin Hartshorne's[38] investigations on Euclidean geometry, so it will be reported below with the name *EPH* (Euclid-von Plato-Hartshorne).

The basic structure of next axiomatization contains a choice of constructive basic concept, the general properties that these basic concepts enjoy, the realization of some ideal situations by means of construction postulates, the properties and uniqueness of constructed objects, and compatibility among the various concepts and constructions.[39] The formal theory under this axiomatization is the *Intuitionistic Type Theory*.

Basic sets

Point: Set
Line: Set
Angle: Set
Circle: Set
Rectilineal Figure: Set

---

[35] You can see also Pambuccian 2008.
[36] Today there are several studies on this topic, but all developed independently of each other. In particular see Hartshorne 2000, Miller 2008, Mumma 2006, Mumma, Avigad and Dean 2009. A study that compares these different approaches will be presented in my Graziani 2013.
[37] von Plato (see bibliography).
[38] Hartshorne 2000.
[39] von Plato (see bibliography).

Basic relations[40]

DiPt (*a,b*) ▷ *a* and *b* are distinct points;
DiL(*l,m*) ▷ *l* and *m* are distinct lines;
DiC(c^1, c^2) ▷ c^1 and c^2 are distinct circles;
L-Apt(*a,l*) ▷ point *a* is left-apart from line *l*;
L-Con (*l,m*) ▷ line *l* is left-convergent with line *m*;
UnOrt(*l,m*) ▷ line *l* and line *m* are unorthogonal;
UnDir(*l,m*) ▷ line *l* and line *m* are unequally directed lines;
UnS(*a,b,c,d*) ▷ point *d* is not as far from *c* as *a* from *b*;
UnAns(*an*(*l,m,a*),*an*(*n,o,b*)) ▷ line *o* does not have the same angular relationship with *n* as *m* has with l^{41} (i.e the two angles are not congruent);
UnEqde(F^1, F^2) ▷ rectilineal figure F^1 and rectilineal figure F^2 are equidecomposable;
UnEqcn (F^1, F^2) ▷ rectilineal figure F^1 and rectilineal figure F^2 have not equal content;
ApAn(*a, β*) ▷ point *a* is apart from angle *β*;
ApC(*a,c*) ▷ point *a* is apart from circle's circumference *c*;
OutC(*a,c*) ▷ point *a* is apart from circle *c*;
OutF(F^1, F^2) ▷ rectilineal figure F^1 is not contained in rectilineal figure F^2;
Nol(F^1, F^2) ▷ rectilineal figure F^1 and rectilineal figure F^2 are nonoverlapping;
TgC(c^1, c^2) ▷ circles c^1 and c^2 are tangent to each other;
SeC(c^1, c^2) ▷ circles c^1 e c^2 intersect to each other;
TgL (*l,c*) ▷ line *l* is tangent to circle *c*;
SeL (*l,c*) ▷ line *l* intersects circle *c*.

Basic constructions

ln(*a,b*) ▷ the connecting line of points *a* and *b*;
pt(*l,m*) ▷ the intersection point of lines *l* and *m*;
par(*l,a*) ▷ the parallel to line *l* through point *a*;

---

[40] ▷ indicates the translation of a formal expression into English.
[41] Where it does not generate misunderstanding I will abbreviate *Ans*(*an*(*l,m,a*), *an*(*n,o,b*)) with *Ans*(*l,m,n,o*).

$ort(l,a)$ ▷ the orthogonal to line l through point a;
$rev(l)$ ▷ the reverse of line l;
$est(l)$ ▷ the extension of line l;
$p1(ln(a,b))$ ▷ the first point on line segment $ln(a,b)$;
$p2(ln(a,b))$ ▷ the ultimate point on line segment $ln(a,b)$;
$bet(ln(a,b),a,b)$ ▷ the point which lies between points a and b on line $ln(a,b)$;
$s(ln(a,b),c)$ ▷ the point which is as far from c as a is from b;
$an(l,m,a)$ ▷ the angle formed by radii l, m and point $pt(l,m)$;
$ans(an(l,m,a),n)$ ▷ the line which has the same angular relationship with the line n, as m with l;[42]
$c(l,a)$ ▷ the circle with radius l and center a;
$ptgl(l,c^1)$ ▷ the point at which line l is tangent to circle c^1;
$ptgc(c^1,c^2)$ ▷ the point at which circle c^1 meets circle c^2;
$psel^1(l,c^1)$ ▷ the first (secant) point at which line l and circle c^1 cut each other;
$psel^2(l,c^1)$ ▷ the second (secant) point at which line l and circle c^1 cut each other;
$psec^1(c^1,c^2)$ ▷ the first (secant) point at which circle c^1 and circle c^2 cut each other;
$psec^2(c^1,c^2)$ ▷ the second (secant) point at which circle c^1 and circle c^2 cut each other.

Derived relations

The readings for these relations are:
$Div(l,a,b)$ ▷ the line l divides two points a and b;
$Bf(l,a,b)$ ▷ point a is before point b on line l;
$Bet(l,a,b,c)$ ▷ point b is between points a and c on line l;
$Apt(a,l)$ ▷ point a is apart from line l;
$CTC(c^1,c^2)$ ▷ circles c^1 and c^2 meet;

---

[42] For simplicity and uniformity where it does not create any misunderstanding I will replace $ans(an(l,m,a),n)$ with $ans(l,m,n)$.

$Min(ln(a,b), ln(c,d))$ ▷ the distance between points a and b is less than the distance between points c and d;
$Inopp(l,m)$ ▷ l and m are inoppositely directed lines;
$R\text{-}Apt(a,l)$ ▷ point a is right-apart from line l;
$R\text{-}Con(l,m)$ ▷ line l is right-convergent with line m;
$DiAn(an(n,o,a), an(m,p,b))$ ▷ $an(n,o,a)$ e $an(m,p,b)$ are distinct angles.

Therefore the derived relations are:
$Div(l,a,b) \equiv (L\text{-}Apt(a,l) \wedge R\text{-}Apt(b,l)) \vee (R\text{-}Apt(a,l) \wedge L\text{-}Apt(b,l))$
$Bf(l,a,b) \equiv DiPt(a,b) \wedge Inc(a \cdot b, l) \wedge Dir(l, ln(a,b))$
$Bet(l,a,b,c) \equiv (Bf(l,a,b) \wedge Bf(l,b,c)) \vee (Bf(l,c,b) \wedge Bf(l,b,a))$
$Apt(a,l) \equiv L\text{-}Apt(a,l) \vee R\text{-}Apt(a,l)$
$CTC(c^1, c^2) \equiv SeC(c^1, c^2) \vee TgC(c^1, c^2)$
$CTL(l,c^1) \equiv SeL(l,c^1) \vee TgL(l,c^1)$
$Min(ln(a,b), ln(c,d)) \equiv Bet(ln(c,d),c,e,d) \wedge S(a,b,c,e)$
$Inopp(l,m) \equiv Undir(l,rev(m))$
$R\text{-}Apt(a,l) \equiv L\text{-}Apt(a,rev(l))$
$R\text{-}Con(l,m) \equiv L\text{-}Con(l,rev(m))$
$Apt(a,l) \equiv L\text{-}Apt(a,l) \vee R\text{-}Apt(a,l)$
$Con(l,m) \equiv Undir(l,m) \wedge Inopp(l,m)$
$DiAn(an(n,o,a), an(m,p,b)) \equiv$
$\equiv UnAns(n,o,m,p)) \vee DiPt(a,b)) \vee DiLn(n,m)) \vee DiLn(o,p)$

Negations of the positive constructive relations

$\neg DiPt(a,b) \equiv EqPt(a,b)$ ▷ a and b are coincident points;
$\neg DiLn(l,m) \equiv EqLn(l,m)$ ▷ l and m are coincident lines;
$\neg DiC(c^1, c^2) \equiv EqC(c^1, c^2)$ ▷ c^1 and c^2 are coincident circles;
$\neg Apt(a,l) \equiv Inc(a,l)$ ▷ point a is incident with line l;
$\neg Con(l,m) \equiv Par(l,m)$ ▷ l and m are parallel lines;
$\neg Unort(l,m) \equiv Ort(l,m)$ ▷ l and m are orthogonal lines;
$\neg Undir(l,m) \equiv Dir(l,m)$ ▷ l and m are (equally) directed lines;
$\neg UnS(a,b,c,d) \equiv S(a,b,c,d)$ ▷ point d is as far from c as a from b;
$\neg UnAns(an(l,m,a), an(n,o,b)) \equiv Ans(an(l,m,a), an(n,o,b))$ ▷ the angle between line o and n is the same as the angle between line l and m;
$\neg UnEqde(F^1, F^2) \equiv Eqde(F^1, F^2)$ ▷ rectilineal figures F^1 and F^2 are equidecomposable;

$\neg \mathrm{Un}Eqcn(F^1,F^2) \equiv Eqcn(F^1,F^2)$ ▷ rectilineal figures F^1 and F^2 have equal content;[43]
$\neg ApAn((a,\beta) \equiv InAn(a,\beta)$ ▷ point a is inside angle β;
$\neg ApC(a,c^1) \equiv InC(a,c^1)$ ▷ point a is incident on circumference of circle c^1;
$\neg OutC(a,c^1) \equiv On(a,c^1)$ ▷ point a is inside circle c^1;
$\neg OutF(F^1,F^2) \equiv OnF(F^1,F^2)$ ▷ the rectilineal figure F^1 is inside rectilineal figure F^2;
$\neg Nol(F^1,F^2) \equiv UNol(F^1,F^2)$ ▷ the rectilineal figure F^1 and rectilineal figure F^2 are overlapping;
$\neg CTC(c^1,c^2) \equiv SepC(c^1,c^2)$ ▷ c^1 and c^2 are distinct circles;
$\neg CTL(l, c^1) \equiv SepL(l, c^1)$ ▷ line l and circle c^1 are separate;
$\neg Inopp(l,m) \equiv Opp(l,m)$ ▷ l and m are oppositely directed lines.

Constructions rules and axioms: the incidence

$$\frac{a:Pt \quad b:Pt \quad DiPt(a,b)}{ln(a,b):Line}[L] \qquad \frac{l:Line \quad m:Line \quad Con(l,m)}{pt(l,m):Point}[PT]$$

$$\frac{l:Line \quad a:Point}{par(l,a):Line}[PAR] \qquad \frac{l:Line}{est(l):Line}[EST]$$

$$\frac{l:Line \quad a:Point}{ort(l,a):Line}[ORT] \qquad \frac{l:Line}{rev(l):Line}[REV]$$

$$\frac{l:Line}{p1(l):Point}[P1] \qquad \frac{l:Line}{p2(l):Point}[P2]$$

[L], [EST] and [PAR] express respectively the first, second and fifth Euclid's postulate.

As in Euclid, even these rules are given here in a minimal form that is refined and strengthened by related axioms and theorems.

---

[43] I could also imagine a definition of indirect type:
$Eqcn(F^1,F^2) \equiv Nol(F^1,F^3) \wedge Nol(F^2,F^4) \wedge Eqde(F^3,F^4) \wedge Eqde((F^1,F^3),(F^2,F^4))$.

IV.2.1 (Uniqueness axiom for constructed straight lines and points)
$DiPt(a,b) \land DiLn(l,m) \supset Apt(a\,|\,b,l\,|\,m)$
its more sophisticated version is:
  IV.2.1*
$DiPt(a,b) \land DiLn(l,m) \supset L - Apt(a\,|\,b,l\,|\,m) \lor R - Apt(a\,|\,b,l\,|\,m)$

IV.2.2 (Uniqueness axiom for parallels)
$DiLn(l,m) \supset Apt(a,l\,|\,m) \lor Con(l,m)$
its more sophisticated version is:
  IV.2.2*
$DiPt(a,b) \land L - Apt(a,l) \supset L - Apt(b,l) \lor L - Con(\ln(a,b),l)$

IV.2.3 (Uniqueness axiom for orthogonals)
$DiLn(l,m) \supset Apt(a,l\,|\,m) \lor Unort(l,m)$

IV.2.4 – IV.2.12 (Axioms for the straight line, intersection point, parallel to line, reverse of line, orthogonal line constructions)
$Par(par(l,a),l)$
$Inc(a, par(l,a))$
$Dir(par(l,a),l)$
$DiPt(a,b) \supset Inc(a \cdot b, ln(a,b))$
$Con(a,b) \supset Inc(pt(l,m), l \cdot m)$
$EqLn(l, rev(l))$
$Opp(ln(a,b), ln(b,a))$
$Ort(ort(l,a), l)$
$Inc(a, ort(l,a))$

IV.2.13 and **IV.2.14** (Refinement axioms)
$Undir(l,m) \lor Inopp(l,m)$
$Con(l,m) \supset L - Con(l,m) \lor R - Con(l,m)$

IV.2.15 and **IV.2.16** (Exclusion axioms for the refinements)
$\neg(L - Apt(a,l) \land R - Apt(a,l))$
$\neg(L - Con(l,m) \land R - Con(l,m))$

IV.2.17 (Compatibility axiom)
$Con(l,m) \lor Unort(l,m)$

Construction rules and axioms: betweenness

$$\frac{a:Point \quad b:Point \quad ln(a,b):Line}{bet(ln(a,b),a,b):Point}[TRA]$$

given a point *a*, a point *b* and a straight line *ln(a,b)* the rule allows you to construct a point between point *a* and point *b*. Where it does not generate misunderstanding I will abbreviate *bet(ln(a,b),a,b)* with *bet(a,b)*.

IV.2.18 (Axiom for constructed point "that lies between")
$Bet(ln(a,c),a,bet(ln(a,c),a,c),c)$

Construction rules and axioms: congruence for line segments

$$\frac{ln(a,b):Line \quad c:Point}{s(a,b,c):Point}[SLN]$$

The rule enables us to find a point *d* which is as far from *c* as *a* from *b*. It enables us to translate segments.

IV.2.19 (Axiom for the construction of point $s(ln(a,b),c)$)
$S(a,b,c,s(ln(a,b),c))$

IV.2.20 (First Euclidean Common Notion)
$S(a,b,c,d) \wedge S(a,b,f,g) \supset S(c,d,f,g)$

IV.2.21 (Second Euclidean Common Notion)
$Bet(l,a,b,c) \wedge Bet(m,d,e,f) \wedge S(a,b,d,e) \wedge S(b,c,e,f) \supset S(a,c,d,f)$

IV.2.22 (Fifth Euclidean Common Notion)
$Bet(l,a,b,c) \supset UnS(a,b,a,c)$

Constructions rules and axioms: congruence for angles

$$\frac{l:Line \quad m:Line \quad a:Point \quad Inc(a,l \cdot m) \quad DiLn(l,m)}{an(l,m,a):Angle}[AN]$$

A Structural and Foundational Analysis of Euclid's Plane Geometry

For convenience, where it will not create any misunderstanding, I will indicate the angles with Greek letters.

$$\frac{an(l,m,a) \quad n:Line}{ans(an(l,m,a),n):Line}[ANS]$$

The rule enables us to construct an angle on the straight line n, congruent to angle $an(l,m,a)$, i.e. it enables to construct a straight line that has the same angular relationship with n, as m with l. Therefore it allows the transportation of angles and it is equivalent to proposition I.23 in the *Elements*.

For simplicity and uniformity we will simplify $ans(an(l,m,a),n)$ with $ans(l,m,n)$, and $Ans(an(l,m,a),an(n,o,b))$ with $Ans(l,m,n,o)$.

IV.2.23 (Axiom for the construction $ans(an(l,m,a),n)$)
$Ans(l,m,n,ans(l,m,n))$

IV.2.24
$Ans(l,m,n,o) \wedge Ans(l,m,p,q) \supset Ans(n,o,p,q)$

IV.2.25 (Triangle congruence, SAS)
$S(a,b,d,e) \wedge S(a,c,d,f) \wedge Ans(ln(b,a),ln(a,c),ln(c,d),ln(d,f)) \supset$
$\supset S(b,c,e,f) \wedge Ans(ln(a,b),ln(b,c),ln(d,e),ln(e,f)) \wedge$
$\wedge Ans(ln(a,c),ln(c,b),ln(d,f),ln(f,e))$

Constructions rules and axioms: intersections of circles and straight lines with circles

$$\frac{a:Point \quad l:Line}{c(a,l):Circle}[CER]$$

$$\frac{l:Line \quad c^1:Circle \quad TgL(l,c^1)}{ptgl(l,c^1):Point}[TGL]$$

$$\frac{c^1:Circle \quad c^2:Circle \quad TgC(c^1,c^2)}{ptgc(c^1,c^2):Point}[TGC]$$

$$\frac{l:Line \quad c^1:Circle \quad SeL(l,c^1)}{p\,sel(l,c^1)\underset{1}{:}Point}[SEL1]$$

$$\frac{l:Line \quad c^1:Circle \quad SeL(l,c^1)}{p\,sel(l,c^1)\underset{2}{:}Point}[SEL2]$$

$$\frac{c^2:Circle \quad c^2:Circle \quad SeC(c^1,c^2)}{p\,sec(c^1,c^2)\underset{1}{:}Point}[SEC1]$$

$$\frac{c^2:Circle \quad c^2:Circle \quad SeC(c^1,c^2)}{p\,sec(c^1,c^2)\underset{2}{:}Point}[SEC2]$$

IV.2.26 and **IV.2.27** and **IV.2.28** and **IV.2.29** (Axioms for constructed circle)
$OnC(a,c(a,l))$
$OutC(b,c(a,ln(a,b)))$
$InC(b,c(a,ln(a,b)))$
$UnS(a,b,a,d) \supset DiC(c(a,ln(a,b)),c(a,ln(a,d)))$

IV.2.30 and **IV.2.31** (Exclusion axioms)
$\neg(TgC(c^1,c^2) \wedge SeC(c^1,c^2))$
$\neg(TgL(l,c^1) \wedge SeL(l,c^1))$

IV.2.32 and **IV.2.33** (Conditions for continuity)
$InC(a,c^1) \wedge On(a,c^2) \wedge InC(b,c^1) \wedge OutC(b,c^2) \supset SeC(c^1,c^2)$
$Inc(a,l) \wedge On(a,c^1) \wedge Inc(b,l) \wedge OutC(b,c^1) \supset SeL(est(l),c^1)$

These two axioms express the conditions for continuity analyzed in section three.

IV.2.34 and **IV.2.35** and **IV.2.36** and **IV.2.37** (Axioms for constructed points *ptgc, psec, ptgl* and *psel*)
$TgC(c^1,c^2) \supset InC(ptgc(c^1,c^2),c^1 \cdot c^2)$
$SeC(c^1,c^2) \supset InC(\underset{1}{p}sec(c^1,c^2) \cdot \underset{2}{p}sec(c^1,c^2),c^1 \cdot c^2)$

$TgL(l,c^1) \supset Inc(ptgl(l,c^1),l) \wedge InC(ptgl(l,c^1),c^1)$

$SeL(l,c^1) \supset Inc(\overset{1}{p}sel(l,c^1) \cdot \overset{2}{p}sel(l,c^1),l) \wedge IncC(\overset{1}{p}sel(l,c^1) \cdot \overset{2}{p}sel(l,c^1),c^1)$

IV.2.38 (Uniqueness axiom for $ptg(c^1,c^2)$)
$DiPt(a,b) \wedge DiC(c^1,c^2) \supset SeC(c^1,c^2) \vee ApC(a|b,c^1|c^2)$

IV.2.39 (Uniqueness axiom for $ptgl(l,c^1)$)
$InC(a \cdot b, c^1) \wedge Inc(a \cdot b, l) \supset (SeL(l,c^1) \wedge DiPt(a,b)) \vee (TgL(l,c^1) \wedge EqPt(a,b))$

Constructions rules and axioms: "equal content"

IV.2.40
$Eqde(F^1,F^2) \supset Eqcn(F^1,F^2)$

IV.2.41
$Eqcn(F^1,F^2) \wedge Nol(F^1,F^3) \wedge Nol(F^2,F^3) \supset Eqcn\big((F^1F^2),(F^2F^3)\big)$

This axiom expresses the second Euclidean Common Notion for the "equal content" relation.

Note that with $Eqcn\big((F^1F^2),(F^2F^3)\big)$ I intend to compare the union of figures F^1F^2 with the union of figures F^2F^3.

IV.2.42
$OnF(F^1,F^2) \wedge OnF(F^3,F^4) \wedge Eqcn(F^1,F^3) \wedge Eqcn(F^2,F^4) \wedge Eqcn\big((F^1,F^5),F^2\big) \wedge$
$\wedge Eqcn\big((F^3,F^6),F^4\big) \supset Eqcn(F^5,F^6)$

This axiom expresses the third Euclidean Common Notion for the "equal content" relation.

IV.2.43
$OnF(F^1,F^2) \wedge Eqcn\big((F^1F^3),F^2\big) \supset UnEqcn(F^1,F^2)$

This axiom expresses the well-known de Zolt's axiom, an equivalent of the fifth Euclidean Common Notion for the "equal content" relation.

The EPH system, which consists of sets, relations and constructions that satisfy the above axioms IV.2.1-IV.2.43, defines the Euclid's plane and *it represents the modern formulation of the axiomatic basis for the development of plane geometry in Euclid's Elements because, as it has been demonstrated in a previous analysis,[44] the following theorem holds:*

Theorem IV. 2. 1 *All the propositions of the books I, II, III, IV of the Elements can be proved in the Euclidean plane, and derived from EPH system of axioms and rules.*[45]

As can be seen, the system has a large number of axioms, but this is due to the nature of the system that uses constructions, axioms on constructed objects and uniqueness axioms, where other systems, like the Hilbertian, can use existential propositions. If you consider that all the principles that deal with constructed objects are deeply intuitive, you can see how the number of axioms to remember can be reduced to twenty-six. Obviously the number of axioms continues to be great, but this is a feature of system as it was built. It is possible to reduce the number of axioms, but this would lead us to replace axioms which are clear and simple with more complex constructions.[46]

As an example of the power of the system, I am about to illustrate the solution in *EPH* to problem I.1 of Euclid's *Elements*:[47]

Theorem (Problem) IV.2.2

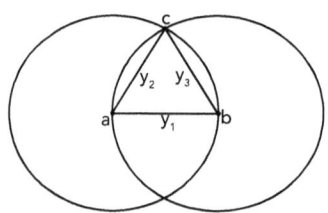

For simplicity: let y_1, y_2, y_3 be given finite straight lines:

---

[44] Graziani 2007, chapter IV.
[45] It is also possible to develop an arithmetic of segments to account for the propositions of books V and VI, but this will not be analyzed here.
[46] Dafa 1997a; 1997b; 1997c; 2003; Dafa, Peifa and Xinxin 2000.
[47] For other theorems you can see Graziani 2007, chapter IV.

Protasis:
$(\exists y_1, y_2, y_3 : Line)$
$\begin{pmatrix} EqPt(p1(y_1), p1(y_2)) \wedge EqPt(p2(y_2), p1(y_3)) \wedge EqPt(p2(y_3), p2(y_1)) \wedge \\ \wedge EqLn(ln(x,w), y_1) \wedge S(p1(y_1), p2(y_1), p1(y_2), p2(y_2)) \wedge \\ \wedge S(p1(y_1), p2(y_1), p2(y_3), p1(y_3)) \wedge \\ \wedge S(p1(y_2), p2(y_2), p2(y_3), p1(y_3)) \end{pmatrix}$
$(ln(x,w) : Line)$

Ekthesis:
$(a,b:Point, \vdash DiPt(a,b), ln(a,b):Line)$

Diorismos:
$(a,b,c : Point, \vdash DiPt(a,b), \vdash Aptl(c, ln(a,b)),$
$\vdash S(a,b,a,c), \vdash S(a,b,b,c), \vdash S(a,c,b,c))$

Kataskeue:
Applying twice the rule [CER] we have $c(a,ln(a,b)):Circle$ and $c(b,ln(b,a)):Circle$.

By construction, point a lies inside the circle $c(a,ln(a,b))$, i.e. $On(a,c(a,ln(a,b)))$. Using the rule [EST] and the Ax. IV.2.33 we can extend the line $ln(a,b)$ up to a point f which lies on $c(b,ln(b,a))$.

By construction point b is between point a and point f, so point f is outside $ln(a,b)$ and $c(a,ln(a,b))$.

But $c(b,ln(b,a))$ has a point inside $c(a,ln(a,b))$ and a point outside $c(a,ln(a,b))$. Using Ax. IV.2.32 two circles meet each other: they are secant.

For the rules [SEC1] and [SEC2] we have two intersection points: we call c the point $\overset{1}{p\,sec}(c^1, c^2) : Punto$. By using the rule [L] we can connect points a,c,b: i.e. $ln(a,c)$, $ln(c,b)$ which together with $ln(a,b)$ constitute the thing sought.

Apodeixis:
It must be proved that the condition holds for the datum and the thing sought built on it. So Euclid proves that segments $ln(a,b)$, $ln(a,c)$, $ln(c,b)$ form an equilateral triangle on $ln(a,b)$.

By Ax. IV.2. 28 and by contraposition of Ax. IV.2.29, since point a is the center of circle $c(a,ln(a,b))$, $ln(a,c)$ is congruent with $ln(a,b)$; and since point b is the center of circle $c(b,ln(b,a))$, $ln(b,c)$ is congruent with $ln(b,a)$.

That is to say, $S(a,c,a,b)$ and $S(b,c,b,a)$.

From Ax. IV.2.9 and Ax. IV.2.20 we obtain $S(c,a,a,b)$ and $S(c,b,a,b)$ and $S(c,a,c,b)$.

Sumperasma:

$(ln(a,b), ln(a,c), ln(c,b) : Line)$

$$\begin{pmatrix} EqPt(p1(ln(a,b)), p1(ln(a,c))) \wedge EqPt(p2(ln(a,c)), p1(ln(c,b))) \wedge \\ EqPt(p2(ln(c,b)), p2(ln(a,b))) \wedge \\ \wedge S(p1(ln(a,b)), p2(ln(a,b)), p1(ln(a,c)), p2(ln(a,c))) \wedge \\ \wedge S(p1(ln(a,b)), p2(ln(a,b)), p2(ln(c,b)), p1(ln(c,b))) \wedge \\ \wedge S(p1(ln(a,c)), p2(ln(a,c)), p2(ln(c,b)), p1(ln(c,b))) \end{pmatrix}$$

$(ln(a,b) : Line)$

Q.E.F.

4.3 A sequent calculus for Euclid's plane geometry

By following contemporary developments of *Substructural Proof Theory* by Jan von Plato and Sara Negri[48] it is also possible formalize a sequent calculus for Euclid's Plane Geometry.[49] I conclude this paper with a brief presentation of this calculus.

For this calculus it is possible to prove[50] the following metatheorems:

Theorem IV.3.1 *The Structural Rule are admissible in extension of G3 sequent system with rules for EPH.*

---

[48] Negri and von Plato 2001; 2011.
[49] This calculation has been presented for the first time in Graziani 2007, chapter IV.
[50] See Graziani 2007, chapter IV.

Theorem IV.3.2 *If $\Gamma \Rightarrow \Delta$ is derivable in the extension of G3 sequent system with rules for EPH, then all formulas in the derivation are either subformulas of the endsequent or atomic formulas.*

In *EPH* sequent calculus there are only two non-mathematical rules: the axioms with zero premises

$$\frac{}{P, \Gamma \Rightarrow \Delta, P}[ass] \qquad \frac{}{\bot, \Gamma \Rightarrow \Delta}(efq)$$

The construction rules are treated as specified in previous paragraph.

The geometrical rules are:

$$\frac{Apt(a,l),\Gamma \Rightarrow \Delta \quad Apt(a,m),\Gamma \Rightarrow \Delta \quad Aptl(b,l),\Gamma \Rightarrow \Delta \quad Apt(b,m),\Gamma \Rightarrow \Delta}{DiPt(a,b),DiLn(l,m),\Gamma \Rightarrow \Delta}[IV.2.1]$$

$$\frac{Apt(a,l),\Gamma \Rightarrow \Delta \quad Apt(a,m),\Gamma \Rightarrow \Delta \quad Con(l,m),\Gamma \Rightarrow \Delta}{DiLn(l,m),\Gamma \Rightarrow \Delta}[IV.2.2]$$

$$\frac{Apt(a,l),\Gamma \Rightarrow \Delta \quad Apt(a,m),\Gamma \Rightarrow \Delta \quad Unort(l,m),\Gamma \Rightarrow \Delta}{DiLn(l,m),\Gamma \Rightarrow \Delta}[IV.2.3]$$

$$\frac{Par(par(l,a),l),\Gamma \Rightarrow \Delta}{\Gamma \Rightarrow \Delta}[IV.2.4]$$

$$\frac{Inc(a,par(l,a)),\Gamma \Rightarrow \Delta}{\Gamma \Rightarrow \Delta}[IV.2.5]$$

A Structural and Foundational Analysis of Euclid's Plane Geometry

$$\frac{Dir(par(l,a),l),\Gamma \Rightarrow \Delta}{\Gamma \Rightarrow \Delta}[IV.2.6]$$

$$\frac{Inc(a,ln(a,b)),\Gamma \Rightarrow \Delta \quad Inc(b,ln(a,b)),\Gamma \Rightarrow \Delta}{DiPt(a,b),\Gamma \Rightarrow \Delta}[IV.2.7]$$

$$\frac{Inc(pt(l,m),l),\Gamma \Rightarrow \Delta \quad Inc(pt(l,m),m),\Gamma \Rightarrow \Delta}{Con(a,b),\Gamma \Rightarrow \Delta}[IV.2.8]$$

$$\frac{EqLn(l,rev(l)),\Gamma \Rightarrow \Delta}{\Gamma \Rightarrow \Delta}[IV.2.9]$$

$$\frac{Opp(ln(a,b),ln(b,a)),\Gamma \Rightarrow \Delta}{\Gamma \Rightarrow \Delta}[IV.2.10]$$

$$\frac{Ort(ort(l,a),l),\Gamma \Rightarrow \Delta}{\Gamma \Rightarrow \Delta}[IV.2.11]$$

$$\frac{Inc(a, ort(l,a)), \Gamma \Rightarrow \Delta}{\Gamma \Rightarrow \Delta}[IV.2.12]$$

$$\frac{L - Apt(a,l), R - Apt(a,l), \Gamma \Rightarrow \Delta}{}[IV.2.13]$$

$$\frac{L - Con(l,m), R - Con(l,m), \Gamma \Rightarrow \Delta}{}[IV.2.14]$$

$$\frac{Undir(l,m), \Gamma \Rightarrow \Delta \quad Inopp(l,m), \Gamma \Rightarrow \Delta}{\Gamma \Rightarrow \Delta}[IV.2.15]$$

$$\frac{L - Con(l,m), \Gamma \Rightarrow \Delta \quad R - Con(l,m), \Gamma \Rightarrow \Delta}{\Gamma \Rightarrow \Delta}[IV.2.16]$$

$$\frac{Con(l,m), \Gamma \Rightarrow \Delta \quad Unort(l,m), \Gamma \Rightarrow \Delta}{\Gamma \Rightarrow \Delta}[IV.2.17]$$

$$\frac{Bet(ln(a,c),a,bet(ln(a,c),a,c),c),\Gamma \Rightarrow \Delta}{\Gamma \Rightarrow \Delta}[IV.2.18]$$

$$\frac{S(a,b,c,s(ln(a,b),c),\Gamma \Rightarrow \Delta}{\Gamma \Rightarrow \Delta}[IV.2.19]$$

$$\frac{S(c,d,f,g),\Gamma \Rightarrow \Delta}{S(a,b,c,d),S(a,b,f,g),\Gamma \Rightarrow \Delta}[IV.2.20]$$

$$\frac{S(a,c,d,f),\Gamma \Rightarrow \Delta}{Bet(l,a,b,c),Bet(m,d,e,f),S(a,b,d,e),S(b,c,e,f),\Gamma \Rightarrow \Delta}[IV.2.21]$$

$$\frac{UnS(a,b,a,c),\Gamma \Rightarrow \Delta}{Bet(l,a,b,c),\Gamma \Rightarrow \Delta}[IV.2.22]$$

$$\frac{Ans(l,m,n,ans(l,m,n)),\Gamma \Rightarrow \Delta}{\Gamma \Rightarrow \Delta}[IV.2.23]$$

$$\frac{Ans(n,o,p,q),\Gamma \Rightarrow \Delta}{Ans(l,m,n,o),Ans(l,m,p,q),\Gamma \Rightarrow \Delta}[IV.2.24]$$

$$\frac{S(b,c,e,f),Ans(ln(a,b),ln(b,c),ln(d,e),ln(e,f)),Ans(ln(a,c),ln(c,b),ln(d,f),ln(f,e)),\Gamma \Rightarrow \Delta}{S(a,b,d,e),S(a,c,d,f),Ans(ln(b,a),ln(a,c),ln(c,d),ln(d,f)),\Gamma \Rightarrow \Delta}[IV.2.25]$$

$$\frac{OnC(a,c(a,l)),\Gamma \Rightarrow \Delta}{\Gamma \Rightarrow \Delta}[IV.2.26]$$

$$\frac{OutC(b,c(a,ln(a,b)),\Gamma \Rightarrow \Delta}{\Gamma \Rightarrow \Delta}[IV.2.27]$$

$$\frac{InC(b,c(a,ln(a,b)),\Gamma \Rightarrow \Delta}{\Gamma \Rightarrow \Delta}[IV.2.28]$$

$$\frac{DiC(c(a,ln(a,b),c(a,ln(a,d)),\Gamma \Rightarrow \Delta}{UnS(a,b,a,d),\Gamma \Rightarrow \Delta}[IV.2.29]$$

$$\frac{}{TgC(c^1,c^2), SeC(c^1,c^2), \Gamma \Rightarrow \Delta}[IV.2.30]$$

$$\frac{}{TgL(l,c^1), SeL(l,c^1), \Gamma \Rightarrow \Delta}[IV.2.31]$$

$$\frac{SeC(c^1,c^2), \Gamma \Rightarrow \Delta}{InC(a,c^1), On(a,c^2), InC(b,c^1), OutC(b,c^2), \Gamma \Rightarrow \Delta}[IV.2.32]$$

$$\frac{SeL(est(l),c^1), \Gamma \Rightarrow \Delta}{Inc(a,l), On(a,c^1), Inc(b,l), OutC(b,c^1), \Gamma \Rightarrow \Delta}[IV.2.33]$$

$$\frac{InC(ptgc(c^1,c^2),c^1), InC(ptgc(c^1,c^2),c^2), \Gamma \Rightarrow \Delta}{TgC(c^1,c^2), \Gamma \Rightarrow \Delta}[IV.2.34]$$

$$\frac{InC(psec(c^1,c^2),c^1), InC(psec(c^1,c^2),c^2), InC(\overset{1}{psec}(c^1,c^2),c^1), InC(\overset{2}{psec}(c^1,c^2),c^2), \Gamma \Rightarrow \Delta}{SeC(c^1,c^2), \Gamma \Rightarrow \Delta}[IV.2.35]$$

$$\frac{Inc(ptgl(l,c^1),l), InC(ptgl(l,c^1),c^1), \Gamma \Rightarrow \Delta}{TgL(l,c^1), \Gamma \Rightarrow \Delta}[IV.2.36]$$

$$\frac{Inc(\overset{1}{psel}(l,c^1),l), Inc(\overset{2}{psel}(l,c^1),l), InCC(psel(l,c^1)c^1), InC(\overset{2}{psel}(l,c^1),c^1)\Gamma \Rightarrow \Delta}{SeL(l,c^1), \Gamma \Rightarrow \Delta}[IV.2.37]$$

$$\frac{SeC(c^1,c^2), \Gamma \Rightarrow \Delta \quad ApC(a,c^1), \Gamma \Rightarrow \Delta \quad ApC(a,c^2), \Gamma \Rightarrow \Delta \quad ApC(b,c^1), \Gamma \Rightarrow \Delta \quad ApC(b,c^2), \Gamma \Rightarrow \Delta}{DiPt(a,b), DiC(c^1,c^2), \Gamma \Rightarrow \Delta}[IV.2.38]$$

$$\frac{SeL(l,c^1), DiPt(a,b)\Gamma \Rightarrow \Delta \quad TgL(l,c^1), EqPt(a,b)\Gamma \Rightarrow \Delta}{InC(a,c^1), InC(b,c^1), Inc(a,l), Inc(b,l), \Gamma \Rightarrow \Delta}[IV.2.39]$$

$$\frac{Eqcn(F^1,F^2),\Gamma \Rightarrow \Delta}{Eqde(F^1,F^2),\Gamma \Rightarrow \Delta}[IV.2.40]$$

$$\frac{Eqcn\big((F^1F^2),(F^2F^3)\big),\Gamma \Rightarrow \Delta}{Eqcn(F^1,F^2),Nol(F^1,F^3),Nol(F^2,F^3),\Gamma \Rightarrow \Delta}[IV.2.41]$$

$$\frac{Eqcn(F^5,F^6),\Gamma \Rightarrow \Delta}{OnF(F^1,F^2),OnF(F^3,F^4),Eqcn(F^1,F^3),Eqcn(F^2,F^4),Eqcn((F^1,F^5),F^2),Eqcn((F^3,F^6),F^4),\Gamma \Rightarrow \Delta}[IV.2.42]$$

$$\frac{UnEqcn(F^1,F^2),\Gamma \Rightarrow \Delta}{OnF(F^1,F^2),Eqcn\big((F^1F^3),F^2\big),\Gamma \Rightarrow \Delta}[IV.2.43]$$

REFERENCES

Acerbi, F. (2011). Perché una dimostrazione matematica greca è generale. *Atti del Workshop "La scienza antica e la sua tradizione"*, eds. F. F. Repellini and E. Nenci. Milan: LED, 25-80.

Blumenthal, L. M. (1961). *A modern view of geometry.* New York: Dover Publications.

Dafa, L. (1997a). Replacing the axioms for connecting lines and intersection points by two single axioms. *Association for Automated Reasoning, Newsletter*, 37, August. URL: <http://www.aarinc.org/Newsletters/037-1997-09.html>

Dafa, L. (1997b). Is von Plato's axiomatization complete for elementary geometry? *Association for Automated Reasoning, Newsletter*, 37, August. URL: < http://www.aarinc.org/Newsletters/037-1997-09.html>

Dafa, L. (1997c). Three axioms of von Plato's axiomatization of constructive projective geometry are not independent. *Association for Automated Reasoning, Newsletter*, 37, August. URL: <http://www.aarinc.org/Newsletters/037-1997-09.html>

Dafa, L., Peifa, J., and Xinxin, L. (2000). Simplifying von Plato's axiomatization of constructive apartness geometry. *Annals of Pure and Applied Logic*, 102, 1-26.

Dafa, L. (2003). Using the Prover ANDP to Simplify Orthogonality. *Annals of Pure and Applied Logic*, 124, 1-3, 49-70.

Feferman, S. (2008). Conceptions of the continuum. URL: <http://math.stanford.edu/~feferman/papers/Continuum-I.pdf>.

Frajese, A. (1950). Sul significato dei postulati euclidei. *Scientia*, vol. LXXXV, N.CDLXIV- 12, 299-305.

Frajese, A. (1968). Il sesto postulato di Euclide. *Periodico di matematiche*, 1-2, 150-159.

Friedman, M. (1985). Kant's theory of geometry. *Philosophical Review*, 94, 455-506. Also in: *Kant and the Exact Sciences*, ed. M. Friedman, 1992. Cambridge: Harvard University Press, 55-95 (I refer to this last edition).

Graziani, P. (2007). *Analisi strutturale e fondazionale della geometria del piano di Euclide.* Urbino: PhD. Thesis, University of Urbino.

Graziani, P. (2013). Contemporary foundations of Euclid's geometry. In preparation for *Isonomia*.

Harari, O. (2003). The concept of Existence and the Role of Constructions in Euclid's Elements. *Archive for History of Exact Sciences*, 57, 1-23.

Hartshorne, R. (2000). *Geometry: Euclid and beyond*. New York: Springer-Verlag.

Heath, T. (1908). *The Thirteen books of Euclid's Elements*. Cambridge: Cambridge University Press.

Hintikka, J., and Remes, U. (1976). *Ancient geometrical analysis and modern logic*, Essay in memory of Imre Lakatos, ed. R. S. Cohen et al., Reidel: Dordrecht, 253-276.

Knorr, W. R. (1983). Construction as existence proof in ancient geometry. *Ancient Philosophy*, 3, 125-148.

Knorr, W. R. (1993). *The ancient tradition of geometrical problems*. New York: Dover Publications.

Mäenpää P. (1993). *The art of analysis. Logic and history of problem solving*. Helsinki: PhD Thesis, University of Helsinki.

Mäenpää, P., and Von Plato, J. (1990). The logic of Euclidean construction procedures, *Language, Knowledge, and Intentionality. Perspectives on the Philosophy of Jaakko Hintikka*, eds. L. Haaparanta, M. Kusch and I. Niiniluoto. *Acta Philosophica Fennica*, 49, 275-293 (The Philosophical Society of Finland, Helsinki).

Martin-Löf, P. (1984). *Intuitionistic Type Theory*. Naples: Bibliopolis.

Miller, N. (2008). *Euclid and his Twentieth Century Rivals: Diagrams in the Logic of Euclidean Geometry*. Stanford: CSLI.

Mumma, J. (2006). *Intuition Formalized: Ancient and Modern Methods of Proof in Elementary Geometry*. Pittsburgh: PhD thesis, Carnegie Mellon University.

Mumma, J., Avigad, J., and Dean, E. (2009). A formal system for Euclid's Elements. *The Review of Symbolic Logic*, 2, 700-768.

Negri, S., and Von Plato, J. (2001). *Structural Proof Theory*. Cambridge: Cambridge University Press.

Negri, S., and Von Plato, J. (2011). *Proof Analysis: A Contribution to Hilbert's Last Problem*. Cambridge: Cambridge University Press.

Norman, J. (2006). *After Euclid. Visual Reasoning and the Epistemology of Diagrams*. Stanford: CSLI Publications.

Panbuccian, V. (2008). Axiomatizing Geometric Constructions. *Journal of Applied Logic*, 6, 24-46.

Panza, M. (2011). Rethinking Geometrical Exactness. *Historia Mathematica*, 38, 42-95.

Panza, M. (2012). The twofold role of diagrams in Euclid's plane geometry. *Synthese*, 186, 1, 55-102.

Proclus (1970). *A Commentary on the First Book of Euclid's Elements*. Princeton: Princeton University Press.

Russell, B. (1917). Mathematics and the Metaphysicians. *Mysticism and Logic* (first published as *Philosophical Essays*, October 1910), ed. B. Russell. London: George Allen & Unwin Ltd.

Sidoli, N. (2004). On the use of term diastema in ancient Greek constructions. *Historia Mathematica*, 31, 2-10.

Sidoli, N., and Saito, K. (2012). Comparative Analysis in Greek Geometry. *Historia Mathematica*, 39, 1-33.

Stenius, E. (1978). Foundations of mathematics: ancient Greek and modern. *Dialectica*, 32, 255-290.

von Plato, J. (1995). Organization and development of a constructive axiomatization. *Types for proofs and programs,* eds. S. Berardi e M. Coppo. Berlin: Springer-Verlag, 288- 295.

von Plato, J. (1997). Formalization of Hilbert's geometry of incidence and parallelism. *Synthese*, 110, 127-141.

von Plato, J. (1998). Constructive theory of ordered affine geometry. *Indagationes Mathematicae*, 9(4), 549-562.

von Plato, J. (2000). *A structural proof theory of geometry*. (Typescript).

von Plato, J. (2001). Terminating proof search in Elementary Geometry. *Istitut Mittag-Leffler, Report* N. 43, 2000/2001.

von Plato, J. (2003). Proofs and Types in Constructive Geometry. (Typescript).

Zeuthen, H. G. (1896). Die geometrische Construction als Existenzbeweis in der antiken Geometrie. *Mathematische Annalen*, 47, 222-228.

CAN THE MATHEMATICAL STRUCTURE OF SPACE BE KNOWN A PRIORI? A TALE OF TWO POSTULATES[*]

EDWIN MARES
Victoria University of Wellington
Edwin.Mares@vuw.ac.nz

ABSTRACT. Michael Friedman has recently revived the Positivists' claim that the geometry of space is conventional, that it is chosen rather than discovered by scientists. He also claims that particular geometries, such as Euclidean geometry and Riemannian geometry, are a priori relative to different physical theories. This paper investigates the more general claim that our mathematical description of space is a priori. In particular, it discusses whether the parallel postulate and the postulate that space has the structure of the continuum are a priori or not. The paper applies different concepts of a priority and determines that there are different answers for different postulates.

1. Introduction

In the epistemology literature, there are various conceptions of what makes a belief a priori. The relationships between these conceptions are often difficult to discern. Some of these conceptions have to do with the way in which beliefs are justified. Some have to do with how beliefs can be refuted. And others, such as nativist theories, have to do with the process of how beliefs or concepts come to be held. In this paper, I consider two important distinctions between types of a priori judgments. The first distinction is between beliefs that are a priori in the *justificational* sense and those that are a priori in the *empirical immunity* sense. The other distinction is between those beliefs that are a priori in an absolute sense and those that are a priori relative to some other beliefs. The notion of the relative a priori has been recently revived by Michael Friedman (especially in Friedman 2001). The central aim of this paper is to examine how keeping these two distinctions clearly in mind affects our judgments about which beliefs in the history of science were (or are) a priori and the sense in which they

---

[*] I would like to thank the participants of the Between Science and Metaphysics conference, especially Alberto Mura (who commented on my paper), Enzo Fano, Francesco Orilia, Massimo Dell'Ultri, and Milosz Arsenijevic.

were. I carry out by discussing the epistemological status of two mathematical postulates and of their application to physical space—Euclid's parallel postulate and the postulate that space is continuous.

The *justificational conception* of a priority says that a belief is a priori if its justification is independent of experience. This is perhaps the most common notion in the literature. Some authors, such as Albert Casullo (2003), hold that it is the central notion of a priority. There is some sense to this, since philosophers often talk about the a priority of a belief as indicating how it is known to be true (that is, we talk of *a priori knowledge*). There are various forms of this conception in the literature. Rationalists, for example, hold that the mind has a faculty of rational insight that can determine independently of experience whether certain propositions are true.

The *empirical immunity* conception of the a priori, on the other hand, says that a belief is a priori if it cannot be refuted by empirical evidence. This notion is also common in the literature. It is the one that Quine discusses in his various attacks on the a priori/a posteriori distinction. The empirical immunity conception is defended by Hartry Field (1998), who thinks that the notion of a priori justification is confused but that the idea of immunity from empirical refutation can be made clear.

In Mares (2011) I follow BonJour (1998) and Casullo (2003) in holding that a belief can be justified independently of experience without being immune to empirical refutation. In this paper I will discuss one form of a priori justification—Aristotelian justification—that I think is particularly appropriate to our knowledge about the geometry of space. This form of a priori justification, I argue, is perfectly compatible with a priori beliefs' being empirically defeasible. My argument for this is in the form of a discussion of the epistemological status of the parallel postulate. It can be justified by Aristotelian means, but it can (and has) also been refuted empirically. So, there is no entailment between the justificational and empirical immunity conceptions of the a priori. Similarly, it seems clear that a belief might be empirically justified but immune to empirical refutation. Various tautologies, such as 'either the sky is blue or it is not blue' can be justified empirically but cannot be refuted by empirical evidence because their negations are contradictions.[1] But the relationship of the relative a priori to each of these other conceptions is not as easy to determine.

---

[1] In fact, I think this is more complicated, since I think that adopting the view that there are true contradictions is a live option, see section 16. But even if we accept that a model of a paraconsistent logic is an accurate representation of the truth, all possible worlds in such a model satisfy all the logical truths of the logic concerned. Those

I also distinguish between the absolute a priori and the relative a priori. A belief is absolutely priori if and only if it is a priori (in either the justificational or empirical immunity sense) regardless of the theories one holds. A belief is a priori relative to a theory if it is *presupposed* by the concepts or propositions in that theory. In section 10 below, I set out a sketch of a formal theory of presupposition that I employ to make the theory of the relative a priori more precise. An intuitive historical example of presupposition in this sense (borrowed from Friedman) is the way in which the concepts and laws of Newtonian physics, such as the notions of acceleration and force, which can hold at an instant of time, depend upon the principles of Newton's calculus. On pre-calculus mathematics, the acceleration of a body could only be described over some non-zero length of time.

In this paper, I delve into this problem of the relationship between the relative a priori and these other two notions of a priority. The relationship between the relative a priori and the empirical immunity conception is particularly difficult. There is no obvious way to show that, in general, if a proposition is a priori relative to a given theory than that proposition cannot be empirically falsified. So, I decided to examine some particular cases that are well-known in the literature on philosophy of science. I have chosen to explore the epistemological status of two mathematical propositions that have been applied to physical space: Euclid's famous parallel postulate and the postulate that space is continuous. I argue that whereas the parallel postulate, as a part of classical physics, can be refuted empirically, it could also previously have been given a priori justification. Moreover, the parallel postulate is not a priori relative to classical physics (contrary to what Reichenbach thought). But the postulate that space is continuous is a priori relative to classical physics and relativity theory. It also can be justified a priori. Moreover, it seems that it cannot be refuted empirically, but this is far from certain.

2. Geometry and the a priori

The advent of non-Euclidean geometries caused a crisis in philosophy and the use of Riemannian geometry in the formalization of general relativity

---

truths may not include the law of excluded middle, but there may be other similar logical truths that can be confirmed empirically.

theory exacerbated this crisis.[2] In the eighteenth century and earlier, philosophers supposed that we had a priori knowledge of the shape of space. This view was most famously held by Leibniz, who held that the principles of Euclidean geometry were analytic, and by Kant, who claimed that we have an a priori intuition of space and that by reflecting on our spatial intuition we can know that space is Euclidean.

The appearance of alternative geometries caused many philosophers to abandon the Kantian position. Bertrand Russell, for one, claims that all the postulates that are not common both to Euclidean and non-Euclidean geometries are empirical, and those that are common are a priori (Russell 1956).

In the 1920s and 30s, the logical positivists—especially Hans Reichenbach—developed a conventionalist view of geometry. Reichenbach argued that we cannot verify any claim that a particular geometry is the one true description of space. It might be, for example, that the instruments that we use to measure distances or the straightness of lines, act differently in different regions of space (Reichenbach 1958). For the positivists, if a scientific view is not verifiable, it must be stipulated, that is, it must be established by convention. And, for the positivists, these sorts of conventions are a priori and analytic.

Michael Friedman has recently interpreted Reichenbach's position as a form of liberalized Kantianism. As Reichenbach pointed out (Reichenbach 1965), there are two elements to Kant's claims that some concept or proposition is a priori. First, there is an absolute notion. That is, a proposition or concept is a priori if it is not revisable. Note that this does not merely mean that it is not prone to empirical refutation; it means that it cannot be undermined for any reason. Second, there is a relative notion of the a priori. As I said in the introduction to this paper, a belief or concept is a priori relative to a theory if the other propositions or concepts of that theory presuppose it. On Reichenbach's view, classical physics presupposes Euclidean geometry and general relativity presupposes Riemannian geometry.

In this paper I look at two mathematical postulates that at times have been applied to physical space: Euclid's fifth axiom and the postulate that

---

[2] I use 'classical physics' here rather than 'Newtonian physics.' The difference is important here. On Newton's own view, light is made of particles which have positive mass. Thus, on Newtonian physics, gravitational forces bend light. Classical physics is the combination of Newtonian mechanics with the wave theory of light and according to it rays of light travel in straight lines unaffected by gravity.

space is continuous. The first of these is, of course, the famous parallel postulate. The second is not part of geometry proper, but that does not matter too much for my present purposes. I am not interested in the epistemological status of geometry per se, but rather I am examining the a priority or a posteriority of mathematical principles as they are applied to the physical world. The form of both of these postulates that I will refer to is the form with which they appear in David Hilbert's *Foundations of Geometry* (Hilbert 1902). Much of my discussion will be about early 20th century science and, because of its influence in that period, it makes sense to refer to that book.

Hilbert presents the parallel postulate not in Euclid's original form, but in the form of Playfair's axiom: "In a plane there can be drawn through any point A, lying outside of a straight line a, one and only one straight line that does not intersect the line a" (Hilbert 1902, 11). This postulate was shown to be independent of the other postulates of Euclidean geometry by J.H. Lambert in 1766 (Torretti 1978, 50). Lambert's argument is the now-standard argument that shows that the surface of a sphere is a model of the other axioms, where straight lines are taken to be great circles on the sphere. Every great circle intersects with every other at some point on a sphere, and so the parallel postulate is falsified on the model.

The axiom stating the continuity of space is stated in two parts in Hilbert. The first part is Archimedes' axiom that says:

> Let A_1 be any point upon a straight line between the arbitrarily chosen points A and B. Take the points $A_2, A_3, A_4, \ldots$ so that A_1 lies between A and A_2, A_2 between A_1 and A_3, A_3 between A_2 and A_4, etc. Moreover, let the segments
>
> $$AA_1, A_1A_2, A_2A_3, A_3A_4, \ldots$$
>
> be equal to one another. Then, among this series of points, there always exists a certain point A_n such that B lies between A and A_n. (Hilbert 1902, 25)

The full mathematical import of Arichimedes' axiom is not of interest to us here. What is important about it is that it ensures that space is infinitely divisible (Boyer 1985, 143).

There is an interesting relationship between Archimedes' axiom and the parallel postulate. Hilbert showed that Archimedes' axiom is independent of the other postulates of geometry. If we drop both the parallel postulate in Hilbert's form and Archimedes' axiom, but retain the postulate that the sum of interior angles of any triangle is 180°, the resulting geometry is consistent with there being infinitely many lines parallel to any line a and point A not on a (Cerroni 2010).

The second part of the axiom of continuity is what Hilbert calls the "axiom of completeness". It says that the geometrical system of planes, lines, and points is complete:

> To a system of points, straight lines, and planes, it is impossible to add other elements in a manner that the system thus generalized shall form a new geometry obeying all five of the groups of axioms. In other words, the elements of geometry form a system which is not susceptible to extension, if we regard the five groups of axioms as valid. (Hilbert 1902, 25)

Hilbert recognizes that this axiom is not usually considered a geometrical postulate, but he wants to ensure that no space (as defined by geometry) has gaps in it. He wants to block the possibility that any extra lines, points, or planes can be added in a geometrical coherent way. He is trying to ensure that the only model for his set of axioms is the "usual Cartesian geometry, obtained when the whole field of real numbers is used in the model" (Corry 2004, 97). In this paper, I will take the intention to accept only spaces based on the reals in this sense as a model for geometry[3], for it is with this that the arguments I consider below take issue.

In what follows, I will discuss, not Archimedes' axiom and the axiom of completeness, but rather the postulate that space is continuous—that the points in space have the order type of the continuum. It does not matter for my present purposes that this postulate is not a postulate of geometry as that field has been construed historically. What is of more interest is the relationship between that postulate and our understanding of physical space.

3. Aristotelian justification

A belief is said to be a priori in the justificational sense if and only if its justification is independent of experience. One of the two theories of justi-

---

[3] I ignore the issue of whether the hyperreals (models for Robinson-style non-standard analysis) or Bell's smooth infinitesimals are more reasonable models for physical space. The issue I deal with is whether space should be seen as continuous or discrete.

fication that I am advancing for our beliefs about physical space is Aristotelianism. I introduce the Aristotelian theory by contrasting it to a more familiar view—rationalism.

Rationalism, according to the definition I give in Mares 2011, is the view that we can know truths by means of some form of rational insight. Historically, philosophers who have called themselves "rationalists" have held that this rational insight is more reliable than empirical methods of belief formation. Some rationalists, such as Descartes and Leibniz, also think that certain concepts are innate. They think, for instance, that the idea of God and the idea of substance are both innate.

Aristotelianism, on the other hand, is the view that we can gain knowledge by means of reflecting on concepts that were gained empirically. We can characterize Aristotelianism as the view that a priori justification (in their sense) requires two steps: (i) that concepts are abstracted from experience and (ii) then these concepts are reflected upon, and this reflection constitutes a form of justification. This form of justification is a priori, because only the reflection on concepts constitutes justification. The abstraction is what could be called an *enabling step*—it enables justification to occur but is not part of the justification.

Aristotelian justification is in some ways similar to the justifying of certain beliefs by appeal to their analyticity. For example, I know that every bachelor is unmarried. I know this because I know the meaning of 'bachelor' in English. I was taught the meaning of 'bachelor' by my parents. I gained an understanding of the word though experience. But by justification for my belief that all bachelors are unmarried is a priori. This is because knowing the meaning of bachelor is sufficient to know that all bachelors are unmarried. No further appeal to experience after learning the meaning of the word 'bachelor' is needed. Now let us consider the case of someone who uses Aristotelian justification to support her belief that nothing can be both red and green all over at the same time. She knows this because she reflects on her concepts of red and green (and the other concepts involved). That she has these concepts and has normal processing abilities is sufficient for her to warrant her belief. She does not need to appeal to induction or other empirical means in her justification.

We can see that the parallel is fairly close. Experience enters the picture in analytic knowledge only through the learning of meanings and in the Aristotelian case only through the acquisition of concepts. It is mental reflection in both cases that extracts the belief from the meaning or the concept and so is solely responsible for the epistemological warrant for the belief.

This does not mean that we cannot turn a case of Aristotelian justification into an empirical justification. Consider the following inference to justify the acceptance of the parallel postulate:

> (**P1**) My concept of space includes the parallel postulate.
> (**P2**) This concept was obtained through experience by abstraction.
> (**P3**) Abstraction through experience and reflection is a reliable belief forming process.
> (**P4**) I should accept beliefs formed through reliable processes.
> ───
> ∴ I should believe that physical space obeys the parallel postulate.

The first premise is, of course, a priori. It comes merely from reflection on the nature of one's own concept. But (P2) and, more importantly (P3), are empirical. Thus, the justificatory inference gives empirical rather than a priori support for its conclusion. But we can do the same thing with analytic beliefs:

> My understanding of the meaning of 'bachelor' entails that every bachelor is unmarried.
> My understanding of the meaning of bachelor was acquired in a reliable manner.
> I should accept beliefs acquired in a reliable manner.
> ───
> ∴ I should believe that all bachelors are unmarried.

This inference does provide empirical support for the belief that all bachelors are unmarried. But notice that the conclusion of this inference and the previous one about space are not the proposition that is to be believed, but rather the statement that we should believe this proposition. Our justifications for beliefs usually take place one level down—at the level of the proposition to be believed. Thus, we also have:

> Bachelors are unmarried men.
> ─────────────────────────────
> ∴ All bachelors are unmarried.

and similarly

$$\frac{\text{Space is Euclidean.}}{\therefore \text{Space obeys the parallel postulate.}}$$

The only premise in each of these inferences are a priori. They are justified by reflection. That this reflection is reliable enables it to do the work it has to do in the justification, but the fact that it is reliable need not be a premise in the justification.

4. Objections to Aristotelianism

One might object to Aristotelianism on the basis that it is metaphysically extravagant. It seems committed to the existence of natures, essences, or universals that are abstracted from things or experiences. Aristotle himself may have been committed to the existence of forms, but 'Aristotelianism' in the sense that I am using it here is compatible with the rejection or acceptance of universals, essences, and so on. Abstraction, on my interpretation, may take the form of the mind's recreating a structure in nature. For example, when we remember patterns and are able to recognize other instances of the same pattern, we can be said to have abstracted that pattern. (We find this view in some structuralists in the philosophy of mathematics, e.g., Bigelow 1988.)

Another possible objection says that Aristotelianism makes error impossible in a priori belief. If we abstract structures that are actually in nature, how can we ever be wrong about them? Clearly, we have been wrong about essential natures of things, in particular for our purposes, people have made mistakes about the structure of space. So this is an important problem.

The answer is that mistakes can be made in either process that define Aristotelianism. First, our abstraction mechanisms may not be perfect. They may cause mismatches between what we think is a structure in the world and the actual structure. Perhaps Aristotle himself would have trouble allowing errors to arise in this way, but in my more liberal sense of 'Aristotelianism' it seems like something that would happen reasonably often. Second, our ability to reflect on the abstracted concepts may not be perfect either. We may make mistakes when, say, we remember or play with structures in imagination.

5. Unification

Aristotelianism, in my opinion, provides us with a very good explanation of how people originally came up with and supported Euclidean geometry as a description of physical space. The reason why Euclidean geometry seems intuitive is that it agrees with the concept of space that we abstract from experience.

But Aristotelianism cannot provide us with a complete explanation of justification of our beliefs about space. Some of our beliefs about space are counterintuitive—they do not match our naïve concept of space. Einstein's view is that there is a four-dimensional space time and our division of it into space and time comes about from our being located in a particular frame of reference, and this view, of course, is almost universally accepted amongst physicists. The logical positivists claim that the postulation of spacetime constitutes the establishing of a convention. Rather than being an attempt to reflect the real nature of the universe, it establishes a convenient mathematical framework in which to do physics.

The claim that scientists stipulate the geometry of space is somewhat difficult to accept. The logical positivists accompanied this claim with the view that there is no fact of the matter about the geometry of space. Reichenbach, in particular, argued that the proposition that space has Euclidean geometry (or that it has Riemannian geometry) is not verifiable. Thus, given the verificationist theory of meaning, there is no fact of the matter concerning which geometry accurately describes physical space. This is an even more difficult view to accept.

A more neutral way of thinking about Einstein's postulation of spacetime is as an attempt to *unify and make coherent* a combination of the views of his predecessors, in particular those of Maxwell and Lorentz. Minkowski spacetime provided Einstein with a clear mathematical framework in which to pursue ideas that were first developed by others. This roots the theory of spacetime not only in the tradition in which Einstein was working but also allows him to adopt the motivations and justifications of his predecessors for his own view.

Unification, of this sort, provides the resulting view with a form of justification. The resulting view inherits whatever justified the views that it unifies. The use of certain of the postulates of geometry—other than the fifth—is justified a priori by Aristotelian means and certain of the other postulates of the theory are justified empirically. The fact that the resulting view unifies these other views into a single coherent theory is further *a priori* justification. The fact that it gives a single, consistent framework in order to deal with electromagnetic phenomena constitutes a priori reasons why it is a better framework than previous theories.

6. Contextualism

The doctrine of contextualism is central to this paper. Contextualism, as the term is used here, is the view that the epistemological status of a statement or theory can only be judged in its historical context. Hilary Putnam, who is one of the best-known contextualists, holds that before the advent of non-Euclidean geometries the parallel postulate was justified a priori (Putnam 1979). The idea here is quite simple. In order to formulate a theory of space, one needs a geometry. If only one is available, then one must choose it. This choice is thereby justified, even in an epistemological sense, for the choice moves forward the cause of theory construction and of science in general.

The main difficulty with advancing a form of contextualism is making clear what is meant by an *available* alternative hypothesis. I won't go into depth concerning this problem, since that isn't required here. But I do say the following. I do not want to claim that a hypothesis is available only if someone has explicitly formulated it. It seems that we might know how to construct a theory of a particular nature without ever doing it and that this should count. Similarly, if all the technical tools that are needed for the construction of a particular sort of hypothesis are available, then it too is available.

In section 10 below, I appeal to the contextual nature of a priority in order to formalize the notion of the relative a priori.

7. Empirical immunity

A belief is said to be a priori in the empirical immunity sense if and only if it cannot be refuted empirically. This is the notion of the a priori that we find in Quine's attacks on a priority and in Hartry Field's entitlement theory of a priori knowledge (Field 1998; 2006).

One might object that if one *knows* that a proposition is empirically immune then she has good grounds to believe that proposition.[4] Thus, the

---

[4] This objection was made in conversation by Jeremy Seligman.

objection continues, empirical immunity is a form of a priori justification. But this objection makes the mistake of conflating having a belief that is empirically indefeasible with the much stronger notion of *knowing* that it is empirically indefeasible. It is only required that the proposition be in fact empirically indefeasible, not that the agent know that it is such. Perhaps knowing, or even believing, that a proposition is empirically indefeasible counts as justification for believing that proposition, but either belief or knowledge that a proposition is empirically indefeasible is not required for that proposition to be the content of an a priori belief.[5]

In fact it may be extremely difficult to know whether a given proposition is immune from empirical refutation. To prove this, one would have to know that the proposition has no empirical consequences. From a syntactical point of view, every sentence has infinitely many sentences that logically follow from it. To be able to survey these consequences to check whether any are empirical is often difficult and perhaps in some cases impossible. Even logical systems that are only moderately complicated like first-order logic are undecidable. We do not have the means to sort the formulas of first-order logic into those that are the consequences of A and those that are not, for every proposition A.

Note that just because a belief may be empirically immune does not mean that it is completely indefeasible. It has been common in the history of philosophy to claim that all a priori beliefs were known with certainty, but that view does not follow from empirical immunity. It may be that non-empirical considerations can lead one to abandon a belief that cannot be empirically refuted. For example, consider the current debates among set theorists about the nature of large cardinals. There is no empirical evidence that can decide the issues concerned. The mathematicians involved use a priori arguments entirely to support their views.

There is another difficulty with the empirical immunity conception of the a priori. This is a circularity problem. It might seem that in order to formulate the conception in a non-circular manner, we need a notion of empirical evidence that can be defined independently of our notion of the a priori. This could be difficult. But there is evidence that is uncontroversially empirical. And perhaps we can formulate the immunity conception to say that a belief is a priori if it cannot be undercut by an amount of evidence that is uncontroversially empirical.

---

[5] The issue of knowing when a proposition cannot be refuted by empirical evidence is interesting. In section 8 I discuss it further.

8. Quine's master argument

Quine's famous arguments against the a priori/a posteriori distinction are largely arguments against the empirical immunity conception. In Mares (2011, ch. 6) I examine two of Quine's arguments very closely. I will look one of these briefly here so that I can extract some lessons regarding the empirical immunity conception that I can use to investigate the epistemology of physical geometry.[6]

This argument, which I call the "master argument," purports to show that no belief is empirically immune. Two central principles that Quine uses in this argument, I have called the *derivation principle* and the *free choice principle*. According to the derivation principle, all of our beliefs can be used to derive empirical consequences (see, e.g., Quine 1966). This adoption of this principle shows the influence of pragmatists like Peirce, James, and Dewey on Quine. They all think that the purpose of beliefs is to have consequences in terms of prediction and action. I won't argue against this principle, because the sorts of beliefs that we are talking about are used as a basis for empirical theories. Surely scientists would not adopt a statement or theory unless they could use it in combination with other theories or statements in order to make empirical predictions.

The free-choice principle can be explained as follows. Suppose that one believes a set of statements Γ that entails a proposition A. Further suppose that she then discovers by experience that $\neg A$. What she should do is contract Γ to a Γ' that is a subset of Γ such that $\Gamma' \cup \{\neg A\}$ is consistent. The free choice principle says that the agent has a free choice (in principle) about which statements in Γ to reject (see, e.g., Quine 1986, 619-620). The parenthetical qualification "in principle" tells us that in practice the choice is often obvious. But the free-choice principle tells us that there is no theoretical bar to rejecting any particular statement.

We can easily see how the derivation and free-choice principles can be used together to produce an argument for the thesis that all of our beliefs are empirically defeasible. Consider an arbitrary statement B that someone believes is true. Then, according to the derivation principle B can be used in combination with other beliefs that the agent holds in order to derive some observation statement, O. But, suppose that O is false, i.e. $\neg O$ is true. Then, by the free-choice principle, the agent can reject B. So, B is not immune to empirical refutation. Generalizing, every belief can be em-

---

[6] For another critique of the immunity conception of a priority, see Kitcher (1985, chs. 1 and 2).

pirically refuted if it can be used together with others in any way to derive an observation statement. This is Quine's master argument.

9. *The relative a priori*

Michael Friedman follows Reichenbach in distinguishing between two elements of Kant's theory of the a priori. Kant's categories are supposed to be absolutely a priori, in the sense that we can be certain of their application a priori and they cannot be refuted by empirical experience. On the other hand, the categories are a priori in a relative sense. In this sense they make possible experiences or beliefs of certain kinds (Friedman 2001, 30). A Kantian example of the relative a priori is the law of cause and effect that, according to Kant, makes it possible for us to give a temporal order to our experiences (Kant 1929, A190/B233ff).

One of Friedman's examples of the relative a priori is the relationship of the calculus to Newtonian mechanics. We could not make sense of Newton's talk of instantaneous acceleration without *some device like* the calculus. On more traditional mathematics, we cannot talk about acceleration over a temporal duration of length zero. Calculus allows us to do so. Friedman says:

> [...] the mathematics of the calculus does not function simply as one more element in a larger conjunction, but rather as a necessary proposition without which the rest of the putative conjunction [of the statements of Newtonian physics] has no meaning or truth-value at all. The mathematical part of Newton's theory therefore supplies elements of the language or conceptual framework, we might say, within which the rest of the theory is formulated. (Friedman 2001, 30)

Friedman uses the fact that one part of a theory may enable another part to function to avoid Quine's master argument. If observation did not fit with Newtonian physics, it was not an option to maintain the theory of the relationship between mass and acceleration, say, and reject the methods of the calculus, for the latter gives the former its meaning. Thus, one does not have a completely free choice as to which proposition to reject.

In the following section, I construct a formalization of the relative a priori in order to provide a tool to analyse whether we can construe statements in certain theories relative to other parts of those theories.

10. Formalising the relative a priori

Let me make the notion of the relative a priori a bit more precise. According to Friedman, the relationship of a statement to one that is a priori relative to it is very like the relationship between a statement and another statement that it semantically presupposes. He says:

> We have said that constitutive principles are necessary conditions of the possibility of properly empirical laws. But this does not mean that they are necessary conditions in the standard sense, where A is a necessary condition of B simply if B implies A. To say that A is a constitutive condition of B rather means that A is a necessary condition, not simply of the truth of B, but of B's meaningfulness or possession of a truth value. It means, in now relatively familiar terminology, that A is a *presupposition* of B. Thus, in the well-known example, originally due to Russell, "The present King of France is bald" presupposes that there is one and only one present King of France, in the sense that the proposition in question lacks a truth value if its accompanying presupposition does not hold. (Friedman 2001, 74)

There are some serious problems with this formulation. First, with regard to some key examples, the historical facts do not seem well described as a form of semantic presupposition. Consider the case of the use of the calculus to formulate Newtonian physics. There is not just one theory but rather a family of theories that allow us to make sense of instantaneous change: Newton and Leibniz's versions of the calculus that use infinitesimals, Weierstrass' version that does not use infinitesimals, Robinson's nonstandard analysis, and Bell's theory of smooth infinitesimals. The fact that there is more than one version of calculus that could be used to formulate the theory makes very complicated the relationship between the theory and the versions of the calculus that were actually used to formulate classical physics. It is therefore difficult to say that Newtonian physics presupposed Newton's calculus in a semantical sense. The statements of Newton's empirical laws could be understood given any of these mathematical theories as background.

Moreover, as Chris Pincock points out (Pincock 2007), it is hard to make sense of the idea that a mathematical theory is presupposed in this manner. Mathematical theories, if true, are necessarily true. So, unless we adopt a theory of semantic presupposition that includes impossible worlds (worlds at which, say, the calculus fails to be true), it would seem that there is a problem with considering what the truth value of a statement would be if a given mathematical theory were false.

I suggest that we can avoid these difficulties by recasting the notion of a presupposition in cognitive terms. It seems that B presupposes A for an agent in the relevant sense if the agent can only make sense of a statement saying that B if he or she believes that A. Let me spell this out more fully. An agent has some theories, methods, and potential beliefs to choose from. These are given historically. For example, in the 18th Century, one could choose to accept Newton or Leibniz's versions of the calculus, but not Weierstrass'. Having chosen one of these, the agent could make sense of the idea that a body has an acceleration at an instant, and so make sense of Newton's theory of forces.

Thus, what we need to model the theory of the relative a priori is a way of modelling cognitive resources. I won't attempt a full formal theory here, but instead will provide a very brief sketch of how I would construct such a theory.

I accept the semantic view of theories, due to Fred Suppe (1989), Bas van Fraassen (1980), and others, and make some slight modifications to it.

I begin with a set of model structures, MS. A model structure is what is usually called a model in the philosophy of science literature. It includes a set of entities and a set of relations, functions, operations, and so on, on that set of entities. I also employ a set of languages, $Lang$. A language $\mathcal{L}$ is an ordered pair (L, V), where L is a set (of the well-formed formulas of $\mathcal{L}$) and V is a *partial* function from formulas and model structures into $\{0,1\}$. A proposition is a pair of sets of model structures. The proposition expressed by a formula A in a language $\mathcal{L}$ is a pair (π^+, π^-), such that π^+ is the set of model structures $\mathcal{M}$ such that $V(A, \mathcal{M}) = 1$ and π^- is the set of model structures $\mathcal{M}'$ such that $V(A, \mathcal{M}') = 0$.

My model of a scientific context at a particular historical time is the pair $< MS, Lang >$, where the set of languages and model structures is historically given. Thus, languages that include Newton's notation for fluxions is included in the model that represents Newton's historical context, but the languages that include modern calculus notation are not included.

Definition 1 *A formula B presupposes a formula A in a model $< MS, Lang >$ if and only if for all model structures $\mathcal{M} \in MS$ and for all languages in our model $\mathcal{L} = (L, V)$, $V(B, \mathcal{M}) \in \{0,1\}$ only if $V(A, \mathcal{M}) = 1$.*

The statement that a particular body has a acceleration r at a specific time t presupposes in Newton's context his calculus, because without the theorems of the calculus being true in a model structure this statement fails to get a truth value. Given a language that does not include the calculus, of a sort that would have been considered an alternative in Newton's time, no understanding of a body's having a particular acceleration at a moment in time would have made sense.

The notion of presupposition explains the asymmetry between a statement and those statements that it presupposes, but it does not give us a full theory of the relative a priori. Surely a statement can presuppose another empirical statement (just as, in the theory of semantic presupposition, 'the present King of France is bald' presupposes 'The King of France exists', both of which are empirical).

In order to give a theory of the relative a priori, I place a binary relation E on model structures such that $E\mathcal{M}\mathcal{M}'$ if and only if $\mathcal{M}$ and $\mathcal{M}'$ are *empirically equivalent*, that is to say, only if we cannot tell these two model structures apart empirically. A proposition π is *purely empirical in the positive sense* if and only if for every $\mathcal{M} \in \pi^+$ if $E\mathcal{M}\mathcal{M}'$ then $\mathcal{M}' \in \pi^+$. Similarly, s *purely empirical in the negative sense* if and only if for every $\mathcal{M} \in \pi^-$ if $E\mathcal{M}\mathcal{M}'$ then $\mathcal{M}' \in \pi^-$.

There is no model theoretic way of defining a proposition's being a priori in the justificational sense. There semantic necessary condition of being a priori in the justificational sense. Suppose that π is a priori in the justificational sense. Then it would seem that there is at least one model structure in π^+ from each set of empirically equivalent model structures. This only says that π is compatible with any possible empirical data. It does not, however, entail that π can be justified at all, let alone justified a priori.

We can, however, give a semantic characterisation of empirical immunity. A proposition σ is *a priori in the empirical immunity sense* if and only if for every model structure in σ^- there is an empirically equivalent model structure that is in σ^+. This means that σ is a priori in the empirical immunity sense if and only if there is no empirical evidence that proves that σ is false.

Turning to the relative a priori, I talk about statements rather than propositions. The theory of the relative a priori uses a set of languages in an important way. The fact that there is this variety of languages allows us to talk about a theory presupposing a statement that expresses a necessary truth. The following is a definition of the relative a priori in both the justificational and empirical immunity senses:

Definition 2 *A statement A is* a priori relative to *B in the justificational sense according to a language $\mathcal{L}$ if and only if B presupposes A and A expresses a proposition according to $\mathcal{L}$ that is a priori in the justificational sense. A statement C is* a priori relative to *D in the empirical immunity sense if and only if D presupposes C and C expresses a proposition according to $\mathcal{L}$ that is a priori in the empirical immunity sense.*

A statement A expresses a proposition π according to $\mathcal{L}$ if and only if $\pi^+ = \{\mathcal{M}: V(A, \mathcal{M}) = 1\}$ and $\pi^- = \{\mathcal{M}: V(A, \mathcal{M}) = 0\}$.

We can see that this modified version of the theory of the relative a priori avoids the difficulties I raised regarding Friedman's view of the theory. Given the variety of languages in a model, we can represent the lack or presence of a mathematical theory without considering worlds in which such a theory is false. Moreover, because the set of languages is determined by a period in history, we can make sense of a presupposition being determined in part by historical context. Thus, the view is a form of contextualism (see section 6 above).

11. *Justification and the relative a priori*

Statements that are a priori relative to our current theories can, in some cases, be justified by Aristotelian means. Some, such as the assertion that space is continuous or that it is Euclidean, may have come about from our reflection on our concept of space. But others cannot be justified in this manner. Some of our current scientific postulates are counter-intuitive, but are a priori relative to the theories concerned. The framework of an integrated spacetime that forms the foundation of the theory of relativity is counterintuitive, as are many of the other foundational ideas in modern physics.

According to Friedman, scientists' choice of a geometry in which to understand physical space is conventional. It is usually thought that conventions are analytic, that is, the statements that constitute a convention are analytically true. But the model theoretic framework that I have set out to understand the relative a priori does not sit perfectly with the analyticity of geometry. Consider two empirically equivalent models for physics, which do not agree on the truth of some geometrical axiom. Because these two models differ in the way in which they relate to geometrical statements, I suggest that they are not cognitively equivalent. They contain different cognitive resources for manipulating and understanding the same empirical phenomena. Unless we accept the wildly counterintuitive claim of the logical positivists that only empirical differences are cognitively important, it seems that we should deny the claim that a choice of a geometrical theory is analytic.[7]

Let's think a bit more about the justification of geometrical theories in science. Consider for a moment, the use of four-dimensional Minkowski space-time in special relativity. Einstein's use of this theory was not merely a stipulation, but is an outgrowth of late-nineteenth century theories, such as those of Maxwell and Lorentz. But Einstein took his view that there is a unified spacetime, because it gave him a coherent and mathematically elegant framework in which to understand the empirical evidence. This framework is pragmatically justified—it is useful, so let's use it!

Now, the usefulness of a theory might be an empirical virtue or a super-empirical virtue of it. If it is useful because it explains the empirical phenomena, then it is an empirical virtue. Note that here I am discussing justification. In so doing, I am interested not in whether a theory is the *only* available theory that explains the phenomena, but rather whether it explains them at all. The fact that non-Euclidean geometry helps to explain empirical phenomena means that it has an empirical justification for its use. This justification does not, by itself, prove that non-Euclidean geometry should be used rather than any other theory, but it does provide it with some justification. The justification that a particular theory should be used rather than any other, if the available theories all fit (or can be combined with other theories to fit) the empirical evidence, must also appeal to the super-empirical virtues of the available theories, to their elegance, simplicity, parsimony, intuitiveness, and so on. Thus, whether a choice of a theory can be considered empirical or a priori in the justificational sense depends on the context; it depends on the other theories available at the time.

---

[7] As far as I know, Friedman only claims that geometrical theories are conventional. He does not claim that they are analytic. What I am adding here is the claim that they can be, under certain circumstances, synthetic.

12. The relative a priori and empirical immunity

In section 10 I set out two conceptions of the relative a priori. If a statement A is a priori relative to a statement B, then, given a scientific context (modelled by a set of model structures and a set of languages), the positive component of the proposition expressed by A contains at least one member from every class of empirically equivalent model structures. Thus, if a piece of empirical evidence e is true in any of the model structures in the model of the scientific context, A cannot be empirically refuted by e.

There seems to be a formulation here of what Kuhn calls "normal science." If the empirical evidence is consistent with a given scientific context, the propositions that are a priori in the empirical immunity sense cannot be falsified by it.

Of course a problem emerges if the empirical evidence is not consistent with the context. In this case, science must adopt new model structures and, perhaps, new languages in order to deal with it. This, in model theoretic terms, is revolutionary science. The issue of whether any empirical phenomenon can be truly inconsistent with a given scientific context and whether scientific revolutions are ever really caused by empirical evidence in this way is of course a difficult one, and I will not pursue it here. But we should note that the theory of the relative a priori as I have set it out does not preclude that a priori propositions in the relative sense can be empirically refuted in extreme situations. This is a consequence of the contextualism that I adopt in section 6.

13. The parallel postulate

In this section, I argue that the parallel postulate should not be taken to be a priori relative to classical physics in the empirical immunity sense as classical physics was understood in the early 20th century.

Riemannian geometry can be axiomatized by removing axiom five from Euclidean geometry and replacing it with one that says that given a line l and a point p not on l *there are zero or more* lines parallel to l that pass through p. (Riemann in fact uses Euclidean geometry minus the parallel postulate as a framework for his differential method of doing geometry. Instead of talking about a whole space, he examines bounded regions. This allows him to discuss isolate parts of space in which, say, the curvature is constant (see Bonola 1955, 139-141).

We can construe the scientific context of the early 1900s such that the model structures that make up the model theoretic construal of classical physics are model structures also of Riemannian geometry. In those model structures of Riemannian geometry that are also model structures of Euclidean geometry, the points of Riemannian geometry are points of Euclidean geometry. And the geodesics of Riemannian geometry are the straight lines of Euclidean geometry. And so on. We can take the basic system to be Riemannian geometry and claim that it (or perhaps merely Euclidean geometry without the parallel postulate) determines the meanings of the terms common to both geometries. Then we can view Euclidean geometry as the set of model structures in which exactly one parallel line exists relative to a line and goes through a point not on that line.

If we do think of the set of models for Riemannian geometry (or even a more general system than that) as determining the meanings of geometrical terms, then it seems that none of the concepts of classical physics depend for their meanings on the parallel postulate. We can take the meanings of physical terms to be determined by a class of models in which there are spacetimes in which there are bodies. Those in which the space is Euclidean (and spacetime otherwise obeys the various postulates, stated and unstated, of classical physics) and the bodies behave according to the laws of classical physics are models of classical physics. It does not seem that any of meanings of the physical concepts of classical physics depend on the parallel postulate. The more general geometrical background seems sufficient.

Moreover, it seems that, from the standpoint of disconfirmation, the parallel postulate can be put on the same footing with the empirical propositions of classical physics. In this argument I borrow heavily from Clark Glymour's work on geometry and bootstrapping (Glymour 1977; 1980). On Glymour's view, in confirming statements in a theory, scientists often hold other statements of that theory fixed and then test the statement in question. We can apply this view both to confirmation and disconfirmation. In the case of the 1919 eclipse, we can construe the events such that scientists were holding the other postulates of classical physics fixed, including the statement that light rays travel along straight lines, and then showed that the apparent motion of the star conflicts with the parallel postulate.

14. *The continuity of space and time in classical physics*

Construing the model structures that make up a theory as a subclass of a wider collection that characterizes some more general theory would seem to be a general method that we can use to show that any given statement fails to be a priori relative to the less general theory (in the empirical immunity sense). But we have to be careful how we apply this method, for it does not always give a true representation of the scientific facts.

As case in point is the postulate that space and time are continuous. We could, in some sense, similarly take the model structures on continuous space to be a subclass of a very wide class of models that include discrete spaces. But in modelling classical physics, for instance, this would do violence to how scientists understood the theory. At the conceptual heart of classical physics is the notion of force. Force to be understood as Newton and the physicists following Newton understood it has to act at instants as well as over positive durations of time. Moreover, bodies must have position, velocity, and acceleration at all times. The understanding of how this is possible requires the calculus and the calculus requires continuous time and space. Thus, the model structures that we use to represent classical physics must represent the continuity of space and time as a priori relative to classical physics in the empirical immunity sense. If we remove the postulate that space and time are continuous, we destroy classical physics as classical physicists understood it.

15. *The continuity of space in quantum physics*

In more contemporary physics, the dense ordering of the points of space has come under attack by the physicist Richard Feynman (1967). The problem concerns the treatment of very small phenomena in physics, the radius of which is less than 10^{-14} cm. The equations that treat larger scale phenomena give absurd answers for phenomena smaller than that (for example, in classical particle physics, the energy mass of an electron goes to infinity). Feynman solved the problem by developing (along with others) a procedure called "renormalization." Feynman, however, did not think renormalization is legitimate. At best it is ad hoc and at worst it is inconsistent. He says:

> The shell game that we play [...] is technically called 'renormalization.' But no matter how clever the word, it is still what I would

> call a dippy process! Having to resort to such hocus-pocus has prevented us from proving that the theory of quantum electrodynamics is mathematically self-consistent. It's surprising that the theory still hasn't been proved self-consistent one way or the other by now; I suspect that renormalization is not mathematically legitimate. (Feynman 1990, 128)

Feynman claims that the failure of the reasonable equations to treat very small entities and processes suggests that it is the axiom of continuity that is at fault:

> I rather suspect that the simple ideas of geometry, extended down to infinitely small space, are wrong. (Feynman 1967, 166)

Penelope Maddy reads Feynman as claiming that we should abandon the axiom of continuity and hold that space is discrete (Maddy 1998, 149), and I agree with Maddy's interpretation.

In order to see what sort of thing Feynman might have in mind, consider a theory of discrete space that has been recently put forward by Rafael Sorkin (in order to act as a basis for a theory of quantum gravity). Sorkin's spaces contain points related by a relation, $\leq$, that is a causal relation. $\leq$ is a partial order (reflexive, transitive, and antisymmetric). It is also *locally finite*. This means that for any points c_1 and c_2, there are only finitely many points between c_1 and c_2. Local finiteness entails that for each $\leq$-branch in a Sorkin space, for any point on that branch, there is a next point along that branch. In other words, the ordering is discrete (see Dowker 2006, Penrose 2005, sect 33.1). So, there are scientific proposals that take the fine-grained structure of space to be discrete.

The question is whether the existence of such proposals undermines the empirical immunity of the continuity of space. I think not. Let us consider two models for Sorkin's theory. In one model, $\mathcal{M}$, space is actually locally finite. In another model, $\mathcal{M}'$, space is continuous, but there is a designated collection, D, of points of the space of $\mathcal{M}'$ such that the spaces of $\mathcal{M}$ and $\mathcal{M}'$ each together with their partial orders are isomorphic with one another. Moreover, only the points of D play any physical role in $\mathcal{M}'$. It would seem that $\mathcal{M}$ and $\mathcal{M}'$ are empirically equivalent to one another. It seems that we can construct a continuous model structure to correspond to any locally finite model structure in this way. Once we have the continuous model structure, we can treat it as a model for Sorkin's theory. We take a first-order translation of Sorkin's theory and then translate this by reinterpreting all the quantifiers as restricted quantifiers ranging only over members of D.

As I have said, it is impossible in many cases at least to prove that a proposition is immune to empirical refutation. What I have done here is to argue only that even if we accept a theory such as Sorkin's which treats particles as jumping between discrete points in pace, we can still have empirically equivalent theories in which space is continuous. The fact that the continuity of space plays no role in explaining physical phenomena may give us warrant to reject such models as irrelevant, but this is not an empirical refutation. It is an *a priori* refutation.

16. Renormalization and paraconistent logic

There is another factor to consider in examining the rejection of the postulate that space is continuous. As we saw in the previous section, Feynman considers the problem of the *consistency* of the standard model of the atom and the use of renormalization as a motivation for rejecting the continuity postulate. What if Feynman's model of the atom is actually inconsistent? Does that provide us with an empirical proof that it is false? According to classical logic, inconsistent theories entail every proposition. Thus, they are empirically false. But, with the rise of paraconsistent logics in the past fifty years and their application to problems in the philosophy of science, this view cannot be maintained without substantial justification.

One strategy for dealing with inconsistent theories is to break them down into consistent subtheories and derive theorems from each of these and then carefully combine theorems from each of these. Bryson Brown and Graham Priest have labelled this technique, "chunk and permeate" (Brown and Priest 2004). It would seem that this technique has been used several times in the history of science. Brown and Priest argue that Newton used it in applying his infinitesimal calculus. In some derivations in *Principia*, Newton divides numbers by infinitesimals, thus treating them as non-zero quantities. But at the end of proofs, if left with infinitesimal quantities, he removes them, thus treating them as equal to zero. On their view, the calculation is divided into sections ("chunks") and only certain information is allowed to pass from one section to another. The same technique seems to have been used by Bohr in his original quantum theory. This theory contained the classical theory of electodynamics. On the Bohr-Rutherford model of the atom, electrons orbit the nucleus. According to classical electrodynamics, electrons should lose energy and crash into the nucleus, but Bohr refrained from taking seriously this aspect of the electrodynamic theory. Instead, he only used certain theorems derived from the theory in making predictions about the behaviour of the atom.

The chunk and permeate view, although it provides a historically accurate reconstruction of scientists' reasoning, may seem to some to be rather disappointing. In dividing a theory into subtheories, it shows how an inconsistent theory can be used to predict results, but not how it can be taken to be a realistic model of nature. Nature is a single interconnected system, not a series of loosely connected systems.

For the more realistic philosophers of science, however, there are other paraconsistent theories (see Mortensen 1995). Some such theories have unified models and could, in that sense, be taken to be the bases of realistic models of the world. But, of course, one would have to adopt some form of dialetheism to accept such models. For by their nature they make contradictions true. (But: any refutation of dialetheism would have to be a priori.)

Given the current context, then, any rejection of a theory just because it is inconsistent must be taken to be a priori, not empirical. For, given the current state of mathematics, it is possible to construct inconsistent physical theories.

It might seem that this use of paraconsistent logic is cheating, or at least undermines my defense of the a priori. One might object that using paraconsistent logic we can always create theories in which space is continuous. It might also be *discrete*, but it will at least be continuous. We can eliminate this problem of cheating by saying that a theory that merely adds on the postulate of continuity without its doing any real scientific work does not really contain that postulate. We merely pretend that it does. The working part of the theory is that based on a discrete view of space. This is the obvious answer to this problem and I think it is the right answer.

Before I leave this topic, I should address another similar issue that has to do with my appeal to paraconsistent logic. It may look to the reader as though the use of paraconsistent logic in this context is unwarranted. I have been discussing the possibility of refuting theories. Given dialetheism—the view that there may be true contradictions—finding evidence that is contrary to one's theory is insufficient to motivate the rejection of that theory. But to say this is to give voice to a common misunderstanding. Given a proposition p, according to paraconsistent logicians, there is an important difference between asserting the negation of p, i.e. asserting $\neg p$, and denying p. Denying p is a speech act that is different from asserting the negation of p. On my view, it is incoherent to assert p and deny p, although it can be coherent to assert both p and $\neg p$. One is correct in his or her denial of p if and only if p fails to be true. In contrast, one is correct in his or her assertion of $\neg p$ if and only if p is false. The difference here is that there are models for paraconsistent logics in which both p and $\neg p$ are true. When a scientist fails to find a result predicted by his or her theory, then he or she *denies* that this result occurred. Hence, the fact that paraconsistent logics allow contradictions to be true has no impact on whether a theory can be falsified.

To sum up this discussion of continuity and empirical defeasibility, it would seem that, unless there is no empirically adequate alternative to a theory based on discrete space available, then we should treat the continuity of space as a priori. It would seem that, unless further argument is given, there are empirically adequate theories available to deal with the phenomena that Feynman cites that are based on continuous theories of space. With regard to whether there will be available adequate theories of quantum gravity based on either discrete or continuous space, or only on one of these alternatives, only time will tell.

17. Conclusion

In this paper, I have looked at the old debates about the epistemological status of geometrical postulates utilizing contemporary tools. I distinguished between two general notions of a priority: the justificational a priori and the empirical immunity concept. And I follow Reichenbach and Friedman in distinguishing between the relative and absolute a priori.

I argued that the parallel postulate may have been given a priori justification, in that it accords with the concept of space that we abstract from our experience, but it is not a priori relative even to classical physics in the empirical immunity sense. The postulate that space and time are continuous, on the other hand, is a priori both in the justificational and empirical immunity sense relative to classical physics. Moreover it is unclear that even in the contemporary scientific context, in which discrete space has been sometimes put forward as a framework for the theory of quantum gravity, that we should view the postulate of the continuity of space as empirically defeasible.

The take-home message of all of this is that we need to be careful to distinguish between those elements of the mathematisation of space and time that we take to be a priori, and the sense in which we mean by 'a priori'.

REFERENCES

Bigelow, J. (1988). *The Reality of Numbers: A Physicalist's Philosophy of Mathematics*. Oxford: Oxford University Press.

BonJour, L. (1998). *In Defense of Pure Reason*. Cambridge: Cambridge University Press.

Bonola, R. (1955). *Non-Euclidean Geometry*. New York: Dover.

Boyer, C. B. (1985). *A History of Mathematics*. Princeton: Princeton Univesity Press.

Brown, B., and Priest, G. (2004). Chunk and Permeate, a Paraconsistent Inference Strategy. Part I: The Infinitesimal Calculus. *Journal of Philosophical Logic*, 33, 379-388.

Casullo, A. (2003). *A Priori Justification*. Oxford: Oxford University Press.

Cerroni, C. (2010). Some Models of Geometries After Hilbert's Grundlagen. *Rendiconti di Matematica*, 7, 47-66.

Corry, L. (2004). *David Hilbert and the Axiomatization of Physics (1898-1918)*. Dordrecht: Springer Verlag.

Dowker, F. (2006). Causal Sets as Discrete Spacetime. *Contemporary Physics*, 47, 1-9.

Feynman, R. (1967). *The Character of Physical Law*. Cambridge: MIT Press.

Feynman, R. (1990). *QED: The Strange Theory of Light and Matter*. London: Penguin.

Field, H. (1998). Epistemological Nonfactualism and the A Prioricity of Logic. *Philosophical Studies*, 92, 1-24.

Field, H. (2006). Recent Debates on the A Priori. *Oxford Studies in Epistemology*, eds. T. S. Gendler and J. Hawthorne, Oxford: Oxford University Press, 69-88.

Friedman, M. (2001). *Dynamics of Reason*. Stanford: Centre for the Study of Language and Information.

Glymour, C. (1977). The Epistemology of Geometry. *Noûs*, 11, 227-251.

Glymour, C. (1980). *Theory and Evidence*. Princeton: Princeton University Press.

Hilbert, D. (1902). *The Foundations of Geometry*. La Salle, IL: Open Court.

Kant, I. (1929). *Critique of Pure Reason*. London: Macmillan.

Kitcher, P. (1985). *The Nature of Mathematical Knowledge*. Oxford: Oxford University Press.

Maddy, P. (1998). *Naturalism in Mathematics*. Oxford: Oxford University Press.

Mares, E. (2011). *A Priori*. Durham and Montreal and Kingston: Acumen and McGill-Queens University Press.

Mortensen, C. (1995). *Inconsistent Mathematics*. Berlin: Springer.

Penrose, R. (2005). *The Road to Reality: A Complete Guide to the Laws of the Universe*. New York: Knopf.

Pincock, C. (2007). The Limits of the Relative A Priori. *Soochow Journal of Philosophical Studies*, 16, 51-68.

Putnam, H. (1979). The Philosophy of Physics. In *Mathematics, Matter, and Method: Philosophical Papers, Vol. 1*. 2nd ed. New York: Cambridge University Press, 79-92.

Quine, W. V. O. (1966). *Ways of Paradox and other Essays*. Cambridge: Harvard University Press.

Quine, W. V. O. (1986). Reply to Jules Vuillemin. *The Philosophy of W. V. O. Quine*, eds. L. E. Hahn and P. A. Schilpp, La Salle, IL: Open Court, 619-622.

Reichenbach, H. (1958). *The Philosophy of Space and Time*. New York: Dover.

Reichenbach, H. (1965). *The Theory of Relativity and A Priori Knowledge*. Berkeley: University of California Press.

Russell, B. (1956). *An Essay on the Foundations of Geometry*. New York: Dover.

Suppe, F. (1989). *The Semantic Conception of Theories and Scientific Realism*. Urbana and Chicago: University of Illinois Press.

Torretti, R. (1978). *Philosophy of Geometry from Riemann to Poincaré*. Dordrecht: Reidel.

van Fraassen, B. (1980). *The Scientific Image*. Oxford: Oxford University Press.

GUNKOLOGY AND POINTILISM: TWO MUTUALLY SUPERVENING MODELS OF THE REGION-BASED AND THE POINT-BASED THEORY OF THE INFINITE TWO-DIMENSIONAL CONTINUUM

MILOŠ ARSENIJEVIĆ AND MILOŠ ADŽIĆ
Department of Philosophy, Philosophy Faculty, University of Belgrade
marsenij@f.bg.ac.rs mradzic@f.bg.ac.rs

ABSTRACT.* In a pure Hilbertian manner, the Neo-Aristotelian Region-Based and the Cantorian Point-Based theory of the infinite two-dimensional continuum are formulated in the extended first-order language $L_{\omega_1 \omega_1}$ through a selection of two appropriate sets of axioms. It is proved that the two theories are only trivially different amongst themselves in the sense defined by Arsenijević 2003, where what matters is not the question of the isomorphism of the basic sets of their models but the sameness of their truth-expressive power. Contrary to the received view, according to which the two theories represent interesting alternatives to each other, it turns out that any of them can be used to express all the truths about its basic and supervening entities and their relations as well as all the truths about its rival's basic and supervening entities and their relations. Consequently, from a metaontological point of view, Gunkology and Pointillism, at least at the two-dimensional level, are just mutually supervening models of the two formal theories that have the same truth-expressive power.

1. Introduction

In 2003, Arsenijević introduced the generalized concept of syntactically and semantically trivial differences between formal theories[1], according to which, two theories non-trivially different in the standard sense – because their respective variables cannot range, in any model, over the elements of one and the same basic set – should still be said to be just trivially different if there are two sets of structure preserving translation rules that map one-one, respectively, the infinite set of all the formulae of one of the theories into a set of formulae of the other one and provide, at the same

---

* This article is a result of our investigation within the projects *Logico-epistemological foundation of science and metaphysics* (179067) and *Dynamical systems in nature and society: philosophical and empirical aspects* (179041) supported by the Ministry of Education, Science and Technological Development of the Republic of Serbia.
[1] See Arsenijević (2003).

time, that all the truths about the elements and their relations in any model of one of the two theories are unequivocally expressed as truths about basic elements and their relations in a model of the other theory, and vice versa. What matters in such cases is not the sameness of the elements of the basic sets in a model of two theories but the equivalence of the truth expressiveness of the two theories, that is, the fact that we can use any of the two theories to express all the truths of the other theory. In any of such cases, any model of one of the two theories can be said to supervene on the corresponding model of the other one.

The reason why in such cases the resulting mappings between the two sets of formulae must be Felix Bernstein's mappings[2] from each of the sets into and not onto another set of formulae follows from the fact that, since the variables of the two theories supposedly cannot range over the elements of one and the same basic set, each element of the basic set of a model of one of the theories must be unequivocally associated with more than one elements in the corresponding model of the other theory, and so, each formula of the language into which we translate must have more variables than the corresponding formula of the language from which we translate.

In their two articles[3], Arsenijević and Kapetanović proved that the Cantorian Point-Based and the Aristotelian Stretch-Based system of the infinite linear continuum are just trivially different in the generalized sense. Each point in a model of the Point-Based system can be corresponded to the abutment place of two equivalence classes of abutting stretches in the corresponding model of the Stretch-Based system and represented through a pair of stretches, whereas each stretch in a model of the Stretch-Based system can be corresponded to and represented through a pair of (end-) points in the corresponding model of the Point-Based system.

In the present article, we want to prove that the same result holds for the Point-Based and the Region-Based system of the infinite two-dimensional continuum, so that, contrary to those who believe that the two systems represent interesting alternatives to each other, each of them can be used equally well for expressing all the truths about its own basic elements and their relations as well about its rival's basic elements and their relations. In particular, as we shall see, any of the two theories can be used at will for expressing truths about one-dimensional entities, which, not being the elements in any model of any of the two systems, can still be spoken of as entities superven-

---

[2] See Cantor (1962), p. 450.
[3] See Arsenijević and Kapetanović (2008a) and Arsenijević and Kapetanović (2008b).

ing either on null-dimensional or on two-dimensional entities of an infinite two-dimensional continuous structure. This will be an additional illustration of why what matters is not the sameness of the set of basic elements in a model but the truth expressiveness of the systems concerning all the entities and their relations present in the model explicitly or implicitly.

2. The Region-Based system S_R

For two reasons we shall formulate both the Region-Based system S_R and the Point-Based system S_P in the infinitary language $L_{\omega_1 \omega_1}$. Both reasons have to do with the main point of the paper. First, since the first-order language is not sufficient for the formulation of the axioms defining a continuous structure, we need some stronger language, but we also want to avoid the standard second-order language, since in S_R we want to speak explicitly only about regions and their relations, just as in S_P we want to speak explicitly only about points and their relations. So, the language $L_{\omega_1 \omega_1}$ represents an appropriate and the weakest possible extension of the first-order language in which we may let the variables of S_R range exclusively over the set of regions and the variables of S_P range exclusively over the set of points as two respective universes of discourse.[4] In this way, we straightforwardly get that to speak of other entities, such as lines, will mean to speak directly just of regions and their relations or about points and their relations. The second reason for choosing $L_{\omega_1 \omega_1}$ is that it will enable us to formulate the two sets of translation rules as directly related to regions as the elements of the basic set of S_R and points as the elements of the basic set of S_P without having to mention explicitly any intermediary entities.[5]

Now, though in sketching the Region-Based system we shall proceed in a pure Hilbertian manner[6], so that what regions and their relations are will follow, after all, from the set of all the axioms implicitly defining them, let us say in advance that in any intended model the regions are meant to be circle-like entities or any other entities topologically homeomorphic to them, which should cover an infinite two-dimensional surface without gaps.

Let S_R contain – besides logical constants, $=, \forall$ and $\exists$ – individual variables $a_1, a_2, ..., a_i, ..., b_1, b_2, ..., b_i, ..., c_1, c_2, ..., c_i, ...$ (sometimes also without subscripts) that will supposedly range over an infinite set of regions as the

---
[4] As it is done, for instance, in Hamblin (1969), Hamblin (1971), Needham (1981), Burgess (1982) and Bochman (1990).
[5] Contrary to what is done in standard formalizations based on Set Theory. See, for instance, Munkres (2000), Ch. 2.
[6] See Hilbert (1902), pp. 447 ff.

basic set of the intended model, which will further be specified through the axioms that implicitly define various relations that hold between regions. Let the only non-logical relation symbol be |, which will, analogously to Scheffer's stroke, serve to define all relations that we want to hold between regions. Intuitively, $a \mid b$ says that regions a and b are connected.

For any kind of connection between any two regions, the following three axioms should hold[7], which we shall call *The Connectedness Axioms of S_R*:

(AS_R1) $\forall a\,(a \mid a)$,

(AS_R2) $\forall a \forall b\,(a \mid b \to b \mid a)$,

(AS_R3) $\forall a \forall b\,(\forall c\,(c \mid a \leftrightarrow c \mid b) \to a = b)$.

The non-connectedness between two regions and some specific topologico - mereological kinds of the connection between two regions can be defined in the following way:

- $a \nmid b \Leftrightarrow_{def} \neg a \mid b$, to be read as "$a$ and b are not connected" (see *diagram 1*)

- $a \sqsubseteq b \Leftrightarrow_{def} \forall c\,(c \mid a \to c \mid b)$, to be read as "$a$ is a part of b"

- $a \sqsubset b \Leftrightarrow_{def} a \sqsubseteq b \wedge a \neq b$, to be read as "$a$ is a proper part of b" (see *diagram 2*)

- $a \circ b \Leftrightarrow_{def} \exists c\,(c \sqsubseteq a \wedge c \sqsubseteq b)$, to be read as "$a$ and b overlap" (see *diagram 3*)

- $a \infty b \Leftrightarrow_{def} a \mid b \wedge \neg a \circ b$, to be read as "$a$ and b are externally connected" (see *diagram 4*)

- $a \triangleleft b \Leftrightarrow_{def} a \sqsubset b \wedge \forall c\,(c \infty a \to \neg c \infty b)$, to be read as "$a$ is an internal part of b" (see *diagram 5*)

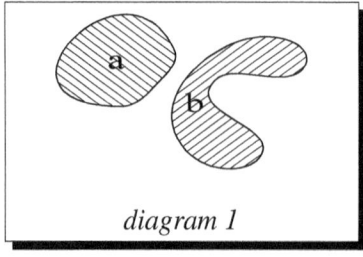

diagram 1

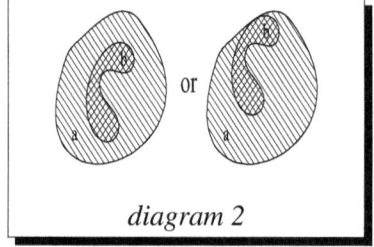

diagram 2

---

[7] An axiomatisation based on the primitive relation of connection is developed in Clarke (1981).

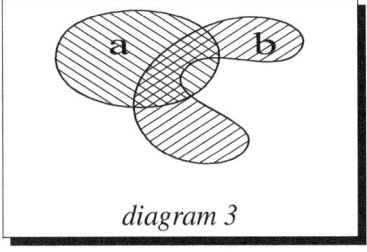
diagram 3

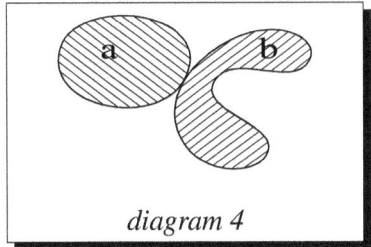
diagram 4

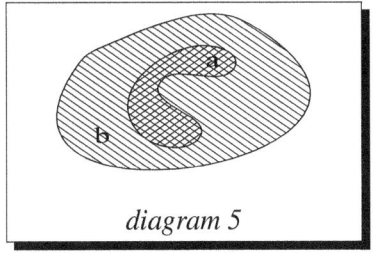
diagram 5

Notice that $\sqsubseteq$ is reflexive, antisymmetric and transitive (defining the partial ordering), whereas $\sqsubset$ and $\triangleleft$ are irreflexive, asymmetric and transitive (defining the strict ordering).

In order to preclude that two disconnected regions make up a region, we shall introduce the following axiom, which we shall call *The Non-Disconnectedness Axiom* (see *diagram 6*, where, if this were not be axiomatically precluded, **a** could be said to be a region consisting of a_1 and a_2):

(AS_R4) $\quad \forall a \forall b \forall c ((b \triangleleft a \wedge c \triangleleft a) \rightarrow \exists d ((d \mid c \wedge d \mid b) \wedge \forall e (e \triangleleft d \rightarrow e \triangleleft a)))$

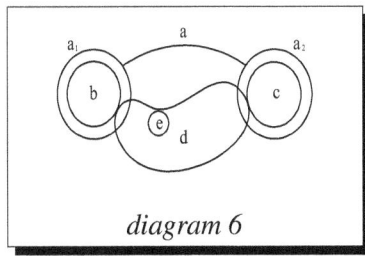
diagram 6

The following axiom, which we shall call *The Gunkness Axiom*, states that every region has proper parts, precluding the existence of atomic regions, which means that, according to the usage of "gunk" that has become popular after David Lewis introduced it[8], the continuum defined by S_R is gunky (see *diagram 7*):

---
[8] See Lewis (1991), pp. 20-21.

(AS_R5) $\forall a \exists b\, (b \triangleleft a)$

Analogously, the existence of a maximal region is precluded by the following axiom, which we shall call *The Inverse Gunkness Axiom* (see *diagram 8*):

(AS_R6) $\forall a \exists b\, (a \triangleleft b)$

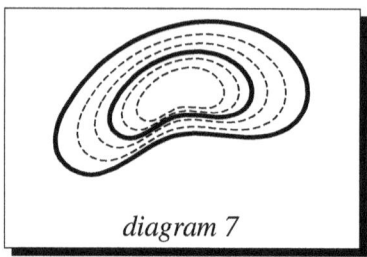
diagram 7

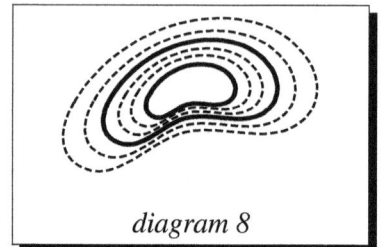
diagram 8

In order to secure that basic elements of S_R be homeomorphic to a circle and not to a two dimensional doughnut, we need the following axiom, which we shall call *The Anti-Torus Axiom* (see *diagram 9*):

(AS_R7) $\forall a \forall b (\forall c (b \triangleleft c \to c \circ a) \to b \triangleleft a)$

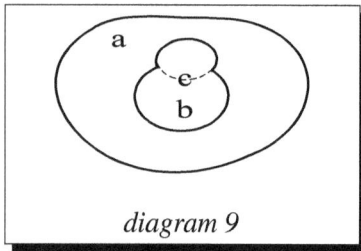
diagram 9

Let us now turn to a technically rather tricky task of defining implicitly the two-dimensionality of the intended model of S_R, which was, according to our best evidence, never done before. Namely, since we want to introduce two-dimensionality in the Hilbertian way, the set of axioms should be so selected as to be interpretable only in the structures that are two-dimensional and not more than two-dimensional. In order to obtain that, we have first, before formulating the two-dimensionality axiom, to define some additional relations in which two or more regions can stand.

To start with, we shall say that the region a is a tangential part of the region b: $a \ll b$, if and only if the following definition is fulfilled (see *diagram 10*):

- $a \ll b \Leftrightarrow_{def} a \sqsubset b \wedge \forall c (a \triangleleft c \to \neg c \sqsubseteq b)$

Gunkology and Pointilism 143

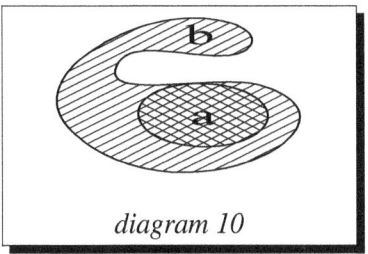

diagram 10

Now, the regions b and c will be said to be (B)ounded by the region a: $B(a,b,c)$, if and only if the following definition is fulfilled (see *diagram 11*):

$$B(a,b,c) \Leftrightarrow_{def} b \ll a \wedge c \ll a \wedge b \infty c \wedge \neg \exists d ((d \infty b \vee d \infty c) \wedge \\ \wedge d \sqsubset a \wedge \forall e (d \sqsubseteq e \wedge e \ll a \rightarrow e \circ b \vee e \circ c)$$

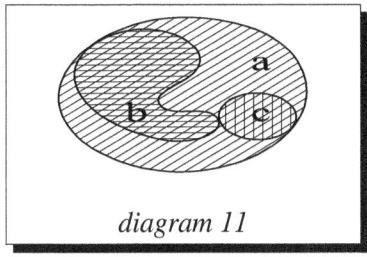

diagram 11

The regions a, b, c, and d will be said to be (N)ested: $N(a,b,c,d)$, if and only if the following definition is fulfilled (see *diagram 12*):

- $N(a,b,c,d) \Leftrightarrow_{def} (a \infty b \wedge a \infty c \wedge a \infty d \wedge b \infty c \wedge b \infty d \wedge c \infty d)$

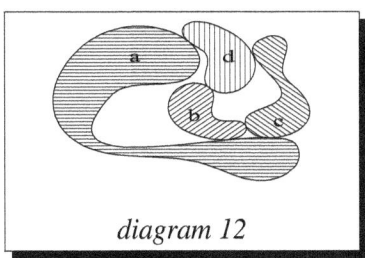

diagram 12

The regions a, b, c, and d will be said to be all mutually externally connected, (W)ittnessing the meeting at a point: $W(a,b,c,d)$, if and only if the following definition is fulfilled (see *diagram 13*):

- $W(a,b,c,d) \Leftrightarrow_{def} \exists e\,(B(e,a,b) \wedge N(a,b,c,d) \wedge c \sqsubseteq e \wedge d \sqsubseteq e \wedge$
  $\wedge \forall f((f \sqsubseteq e \wedge f \mid c \wedge f \mid d) \rightarrow (f \mid a \vee f \mid b)))$

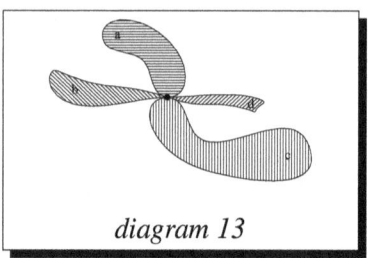

diagram 13

The two regions a and b will be said to be externally connected by meeting at a (P)oint: aPb, if and only if the following definition is fulfilled (see *diagram 14*):

- $aPb \Leftrightarrow_{def} \exists c \exists d\, W(a,b,c,d)$

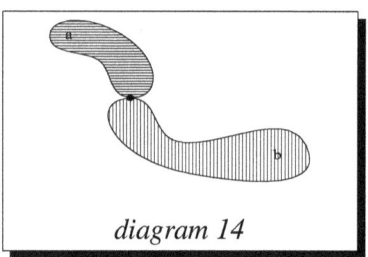

diagram 14

Here is the right place to note that we mention the *point* only informally and intuitively when we speak about regions meeting at a point since in S_R the points are not introduced explicitly. If we wanted, we could introduce them explicitly, but we do not need to do so. We may wait till they occur as elements in the Point-Based system and begin to speak in S_R about them only then, with the use of the translation rules holding between the two systems.

The region a will be said to meet with b and c at the point at which the regions b and c meet: $D(a,b,c)$, if and only if one of the (D)isjuncts from the following definition is true (see *diagram 15*):

- $D(a,b,c) \Leftrightarrow_{def} \begin{array}{l} bPc \wedge (((a \sqsubseteq b \vee b \sqsubseteq a \vee a \circ b) \wedge aPc) \vee \\ \vee ((a \sqsubseteq c \vee c \sqsubseteq a \vee a \circ c) \wedge aPb)) \end{array}$

Gunkology and Pointilism

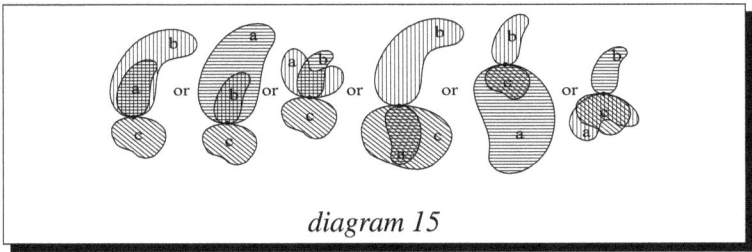

diagram 15

We are now ready to formulate the two-dimensionality axiom. First, it should be noted that the last definition says only when the region a should be said to meet with b and c at the point at which b and c meet but does not preclude the possibility that a meets b and c at the point at which b and c meet even if no disjunct from the definition of $D(a,b,c)$ is true. It is easy to see that a could meet b and c at the point at which b and c meet by approaching the meeting point along some third direction, without having any part common with c and d. But if the relational structure in which S_R is interpreted were two-dimensional, it would not be possible that no disjunct from the definition of $D(a,b,c)$ be true. So, in order to provide that the structure is two-dimensional, we have only to state, by *The Two-Dimensionality Axiom*, that one of the disjuncts is true in any given case:

$(AS_R 8)$ $\quad \forall a,b,c_1,d_1,c_2,d_2(W(a,b,c_1,d_1) \wedge W(a,b,c_2,d_2) \rightarrow$
$\rightarrow D(c_1,c_2,d_1) \wedge D(c_1,c_2,d_2))$

The last thing we have to do is to preclude axiomatically the possibility of "holes" within the set of all regions. This can be done analogously to the way in which Cantor precluded the existence of "gaps" in the one-dimensional continuum. The set of null-dimensional points which is dense is also continuous if any infinite accumulation of points that has a limit is such that the limit is an element of the basic set itself.[9] In our case, any infinite accumulation of regions, where a successive region contains the previous one as its internal part (see *diagram 16* below), has to be such that its limit is a region from the basic set that contains all the accumulating regions as its internal parts. The axiom will be formulated by the use of the following two definitions of $\varphi(a,b_1,b_2,...)$ and $\psi(a,b,c_1,c_2,...)$:

- $\varphi(a,b_1,b_2,...) \Leftrightarrow_{def} \bigwedge_{n<\omega} b_n \triangleleft a \wedge \bigwedge_{m<n<\omega} b_m \triangleleft b_n \wedge$
$\wedge \neg \exists c(\bigwedge_{n<\omega} b_n \triangleleft c \wedge c \triangleleft a)$

- $\psi(a,c,b_1,b_2,...) \Leftrightarrow_{def} c \infty a \wedge \forall d(c \triangleleft d \rightarrow \bigvee_{n<\omega} b_n \circ d)$

---
[9] Cf. Cantor (1962), p. 194.

Now, *The Continuity Axiom* reads as follows:

(AS_R9) $\forall a(\forall b_n)_{n<\omega}(\varphi(a,b_1,b_2,...) \to \exists c\psi(a,c,b_1,b_2,...))$

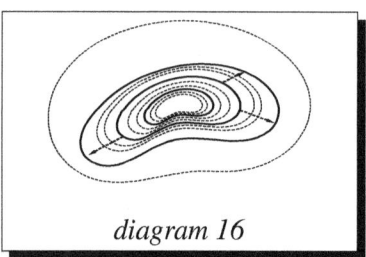

diagram 16

Obviously, the above axiom precludes the existence of a hole *outside* the region a. One may raise the question whether we should formulate an additional axiom which will preclude the existence of a hole *within* a. However, the axiom (AS_R5) together with the axiom (AS_R6) is sufficient to preclude such a possibility. Namely, according to the axiom (AS_R5), there must be a region which is a proper part of a and then, starting from *this* region, we can always apply to it the axiom (AS_R6) which guarantees the absence of the alleged holes within a.

3. The Point-Based system S_P

Intuitively, in order to cover completely an infinite two-dimensional plane (or a two-dimensional surface topologically homeomorphic to it) by a set of null-dimensional points, we need two sets of points, each making up an infinite set of parallels (parallel straight lines in the case of a plane or parallel quasi-straight-lines in the case of a surface topologically homeomorphic to it) such that each point from the first set of parallels is a point common with one and only one point from the other set, and vice versa. In what follows we use *parallels* to refer both to parallel lines and to quasi-parallel lines in the sense explained above (see *diagram 17*). In such a way, the continuity of each line (supervening on the set of points arranged so as to build up the linear continuum) from one of the sets will guarantee the continuous order of the lines of the other set, and vice versa.

Gunkology and Pointilism

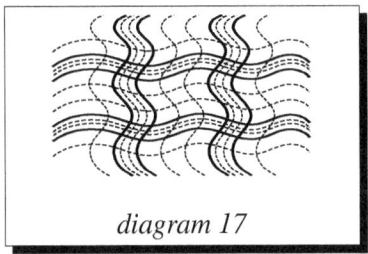

diagram 17

In order to express the required arrangement of points axiomatically, we shall first define $\varphi_1(\vec{\alpha}), ..., \varphi_{10}(\vec{\alpha})$ as shorthands for the formulae which are axioms of the Point-Based system of the infinite linear continuum as they are formulated in $L_{\omega_1\omega_1}$ by Arsenijević and Kapetanović.[10]

1. $\varphi_1(\vec{\alpha}) \Leftrightarrow_{def} \bigwedge_{j<\omega}(\neg \alpha_j < \alpha_j)$
2. $\varphi_2(\vec{\alpha}) \Leftrightarrow_{def} \bigwedge_{j,k,l<\omega}((\alpha_j < \alpha_k \wedge \alpha_k < \alpha_l) \to \alpha_j < \alpha_l)$
3. $\varphi_3(\vec{\alpha}) \Leftrightarrow_{def} \bigwedge_{j,k<\omega}(\alpha_j < \alpha_k \vee \alpha_k < \alpha_j \vee \alpha_j = \alpha_k)$
4. $\varphi_4(\vec{\alpha}) \Leftrightarrow_{def} \bigwedge_{j,k,l<\omega}((\alpha_j = \alpha_k \wedge \alpha_j < \alpha_l) \to \alpha_k < \alpha_l)$
5. $\varphi_5(\vec{\alpha}) \Leftrightarrow_{def} \bigwedge_{j,k,l<\omega}((\alpha_j = \alpha_k \wedge \alpha_l < \alpha_j) \to \alpha_l < \alpha_k)$
6. $\varphi_6(\vec{\alpha}) \Leftrightarrow_{def} \bigwedge_{j<\omega} \alpha_j \bigvee_{k<\omega} \alpha_k(\alpha_k < \alpha_j)$
7. $\varphi_7(\vec{\alpha}) \Leftrightarrow_{def} \bigwedge_{j<\omega} \alpha_j \bigvee_{k<\omega} \alpha_k(\alpha_j < \alpha_k)$
8. $\varphi_8(\vec{\alpha}) \Leftrightarrow_{def} \bigwedge_{j,k<\omega}(\alpha_j < \alpha_k \to \bigvee_{l<\omega}(\alpha_j < \alpha_l \wedge \alpha_l < \alpha_k))$
9. $\varphi_9(\vec{\alpha}) \Leftrightarrow_{def} ((\exists \beta (\bigwedge_{j<\omega} \alpha_j < \beta) \to$
   $\to \exists \gamma (\bigwedge_{j<\omega} \alpha_j < \gamma \wedge \neg \exists \delta (\bigwedge_{j<\omega} \alpha_j < \delta \wedge \delta < \gamma)))$
10. $\varphi_{10}(\vec{\alpha}) \Leftrightarrow_{def} ((\exists \beta (\bigwedge_{j<\omega} \beta < \alpha_j) \to$
    $\to \exists \gamma (\bigwedge_{j<\omega} \gamma < \alpha_j \wedge \neg \exists \delta (\bigwedge_{j<\omega} \delta < \alpha_j \wedge \gamma < \delta)))$

It is important to note why in the above definitions we should and could omit the universal quantification present at the beginning in the corresponding axioms of the Point-Based axiomatization of the linear continuum. On the one hand, we have had to omit the universal quantification because we need the variables to be free in view of the intended definition of a set of points that make up an infinite set of parallel lines. On the other hand, in view of the way in which the above definitions are introduced, the omission of the universal quantification doesn't allow for the possibility of producing

---
[10] See Arsenijević and Kapetanović (2008a) and Arsenijević and Kapetanović (2008b).

counterexamples. Let us suppose, for example, that one wants to produce a counterexample to what is implied by definition 9. For doing this, he could introduce new variables instead of those present in the definition and suppose that in this particular case there is no least upper bound in spite of the fact that there is an upper bound. However, there is nothing that can prevent us to re-introduce systematically the variables that occur in definition 9 instead of the new variables and show in this way that the alleged counterexample is just an instance of what is said in the definition.

Given the previous ten formulae, the following formula defines a set of points standing in such relations that they make up an infinite set of parallels that as such may though need not be continuously ordered:

$$\Psi(\vec{\alpha}^1, \vec{\alpha}^2, \ldots) \Leftrightarrow_{def} \bigwedge_{j \leq 10} \bigwedge_{i < \omega} \varphi_j(\vec{\alpha}^i) \wedge \neg \exists x, y \left(\alpha_n^i = x = y = \alpha_m^j \right),$$
for $i, j, m, n < \omega$ and $i \neq j$

Now, following the intuitive suggestion, the continuity of a set of parallels defined by the last formula will be guaranteed by letting this and some other set of parallels cut in the way intuitively described above. In addition, in order to secure that the relational structure is just two-dimensional, it is necessary not only that there are two sets of parallels that cut in the way required but also that for any third set of parallels it holds that there is no point at them that would not be one of the points from the two sets of cutting parallels. All this will be implicitly defined by the following axiom, which we may call *Descartes' Axiom* (remembering the concept of the *Cartesian Product*, as a result of which Descartes became famous in the history of mathematics[11]):

(AS$_P$1) $(\exists \vec{\alpha}^n)_{n < \omega} (\exists \vec{\beta}^m)_{m < \omega} (\Psi(\vec{\alpha}^1, \vec{\alpha}^2, \ldots) \wedge \Psi(\vec{\beta}^1, \vec{\beta}^2, \ldots) \wedge$
$\wedge \alpha_j^i = \beta_k^l \wedge \neg \alpha_j^i = \beta_q^p \wedge \forall \gamma (\bigvee_{r,s < \omega} \gamma = \alpha_s^r)),$
for $l \neq p, k \neq q$ and $i, j, l, k, p, q < \omega$

It is important to note that, contrary to the Region-Based system, in which we had to preclude the possibility of "holes" by introducing a special axiom, we don't have to do that in the case of the Point-Based system. Namely, though no single infinite set of parallels as such precludes the existence of "holes" within it (see *diagram 18* bellow), *Descartes' Axiom* precludes the possibility that this holds for an infinite set of parallels that are continuous, since the way in which the "cut" of the two sets of parallels is defined and stated to exist by (AS$_P$1) imposes the continuity of all these parallels themselves (see *diagram 17* above). The existence of "holes" within a set of continuous

---
[11] See, for instance, Boyer and Merzbach (2011), p. 319.

parallels is precluded by the non-existence of Cantor "gaps" within any of the cutting parallels.

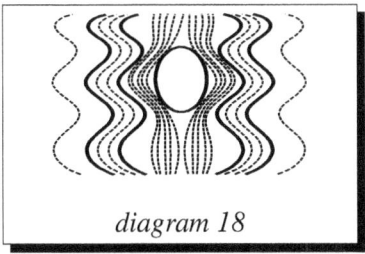

diagram 18

4. Translation rules

4.1. Translation of S_P into S_R

Intuitively, two regions in a model of S_R, a and b, which stand in the meeting-at-a-point relation aPb, meet just at one single point of the corresponding model of S_P. However, there is an infinite number of pairs of other regions that also meet at that very point. So, there is an infinite number of ways in which one and the same point of S_P can be identified within S_R. Fortunately, however, all these ways can be exhaustively classified as only 25 kinds of ways in which two pairs of regions, a and b, and c and d, can be related if a and b, and c and d should meet at one and the same point. Namely, regions a and c can stand in one of the five possible relations of the following kind: $a = c$, $a \sqsubseteq c$, $c \sqsubseteq a$, $a \circ c$, or $a \infty c$, and the same holds *mutatis mutandis* for b and d, where in the case of overlapping, say $a \circ c$, there must be a proper part of a region c which is also a proper part of a and which meets region b at-a-point, and where in the case of $a \infty c$ and $b \infty d$, the said regions must stand in the witnessing-meeting-at-a-point relation $W(a,b,c,d)$.

In order to secure that the identity of two points in S_P when spoken of in S_R is just about the unique *point* we need a function f_1, mapping the variables of S_P into the variable pairs of S_R:

$$f_1 : \alpha_n \rightarrow \langle a_{2n-1}, a_{2n} \rangle \qquad (n = 1, 2, ...)$$

In order to obtain a compact version of the first translation rule T_1, we shall introduce the shorthands for the five possible ways described above in which regions a_{2m-1}, a_{2n-1}, a_{2m} and a_{2n} may be:

- $\Phi_1(A,C) \Leftrightarrow_{def} A = C$
- $\Phi_2(A,C) \Leftrightarrow_{def} A \sqsubseteq C$

- $\Phi_3(A,C) \Leftrightarrow_{def} C \sqsubseteq A$
- $\Phi_4(A,C) \Leftrightarrow_{def} A \circ C \rightarrow \begin{array}{l} \exists x,y((x \sqsubseteq A \vee x \sqsubseteq C \vee y \sqsubseteq A \vee y \sqsubseteq C) \wedge \\ \wedge (xPA \vee xPC \vee yPA \vee yPC)) \end{array}$
- $\Phi_5(A,C) \Leftrightarrow_{def} A \infty C$

where A and C stand for a_{2m-1} and a_{2n-1}, or a_{2m} and a_{2n} respectively.

The identity of α_m and α_n in S_P should be expressed by the formula of S_R that stands on the right side in the translation rule T_1:

$$(T_1) \quad \alpha_m = \alpha_n =^{T_1} APB \wedge CPD \wedge \bigvee_{i,j \leq 5}(\Phi_i(A,C) \wedge \Phi_j(B,D)) \wedge \\ \wedge ((\Phi_5(A,C) \wedge \Phi_5(B,D)) \rightarrow W(A,B,C,D))$$

where A, B, C and D stand for a_{2m-1}, a_{2n-1}, a_{2m} and a_{2n} respectively.

Two things here should be noted. Firstly, in defining the formula Φ_4 we could make use of variables x and y for speaking about proper parts of A and C owing to the fact that the relation defined for A and C is independent of the the way in which the the regions B and D stand (see *diagram 19*).

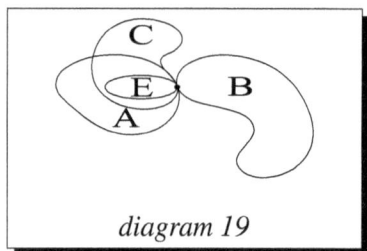

diagram 19

Secondly, we could not rely on the same technique in defining the formula Φ_5 because, unlike the previous case, we need to take into account all the four regions meeting at the same point in order to preclude that the two corresponding pairs meet at different points, and that is why we had to introduce the witnessing-meeting-at-a-point in order to preclude such a possibility. The *diagram 20* illustrates the case that should be precluded:

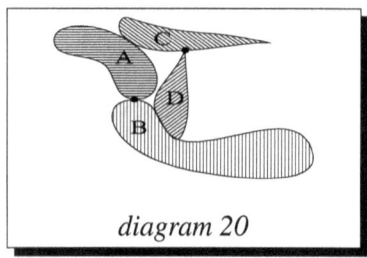

diagram 20

It is important to notice that (T_1) is *not* a *definition* that would make the formulae flanking the $=^{T_1}$ sign interchangeable within any of the two systems, since the formula on the left side of the $=^{T_1}$ sign is not a formula of S_R just as the formula on the right side is not a formula of S_P. What (T_1) enables us to do is only to *express* a truth expressed in S_P as the *corresponding* truth of S_R. "*What matters is the equivalence of the truth expressiveness and not the sameness of the sets of basic elements*".

The second translation rule should tell us how to translate the speaking of a given line within S_P into the speaking of that line within S_R. This case is particularly interesting because the lines are not basic elements neither in a model of S_P nor in a model of S_R. The lines are *supervening* entities both in the models of S_P as well in the models of S_R. So, this case shows *a fortiori* why "*what matters is the equivalence of the truth expressiveness and not the sameness of the sets of basic elements*".

In order to formulate the second translation rule, we have first to formulate both how a line is to be spoken of in S_P as well as how it is to be spoken of in S_R. The crucial difference between these formulations and the translation rule we are looking for consists in the fact that these two formations will be *definitions* that enable us to speak of lines *within* the first and the second system, respectively.

Given the definitions of $\varphi_1, \varphi_2, ..., \varphi_{10}$ above, the speaking of a line within S_P can be easily defined as speaking of the points $\chi(\alpha_1, \alpha_2, ...)$ so that:

$$\chi(\alpha_1, \alpha_2, ...) \Leftrightarrow_{def} \bigwedge_{j \leq 10} \varphi_j(\alpha_1, \alpha_2, ...)$$

For the definition of a line within S_R, we will use the definition of the tangential part $\ll$ introduced above and restrict it in two steps suggested by *diagram 21*.

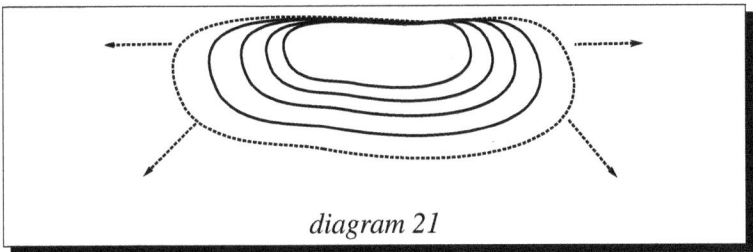

diagram 21

First, the relation $<$, as a restriction of $\ll$, will be defined as a binary relation on the set of all regions so that:

$$a \lessdot b \Leftrightarrow_{def} a \ll b \wedge \neg \forall c ((cPa \wedge \neg c \circ b) \rightarrow cPb)$$

Intuitively, $a \lessdot b$ says that the region a is a tangential part of the region b but so that the two regions share necessarily more than a point on the boundary.

And now, secondly, let $\lessdot^*$ be the *transitive closure* of a relation $\lessdot$ i.e., the smallest transitive relation containing $\lessdot$. This means that any tangential part is developing along an infinite line.

In order to secure that by speaking about a line in S_P and in S_R we are speaking of one and the same line, we need the function f_2 mapping an infinite-tuple of variables of S_P into an infinite-tuple of variables of S_R:

$f_2 : (\alpha_1, \alpha_2, ...) \rightarrow (a_1, a_2, ...)$

So, in view of these two definitions and in view of how the speaking of a line is defined above for the system S_P, the second translation rule T_2 should read as follows:

(T_2) $\qquad \chi(\alpha_1, \alpha_2, ...) =^{T_2} a_1 \lessdot^* a_2 \wedge a_2 \lessdot^* a_3 \wedge ...$

4.2. Translation of S_R into S_P

Intuitively, a *region* within a model of S_P can be understood as a two-dimensional circle, or any other figure homeomorphic to it, consisting completely of a set of points. In what follows we shall refer by *one-dimensional circle* both to a genuine circle-line as well as to any closed line topologically homeomorphic to it, whereas by *two-dimensional circle* we shall refer to a genuine two dimensional circle-region as well as to any region topologically homeomorphic to it. To find out how we can speak in S_P of an entity of the latter kind is, perhaps unexpectedly, a rather tricky task. The hint is to find a way to speak of a *line segment* and of a *closed line*, and then define a one-dimensional circle as a set of points such that it consists of all the segments having two points of the closed line as its end-points and consisting only of the elements from the set of points constituting the one-dimensional circle itself.

Following the hint, the first thing to do is to see when the points $\alpha_1, \alpha_2, ...$ make up a line segment, which we shall denote by $\varsigma(\alpha_1, \alpha_2, ...)$. Remembering the way in which a line is defined within S_P, it becomes clear that we have to leave out $\varphi_6 - \varphi_7$ from $\varphi_1 - \varphi_{10}$, which implicitly define the continuum as infinite, and introduce φ_6^* and φ_7^* instead (see *diagram 22*):

Gunkology and Pointilism

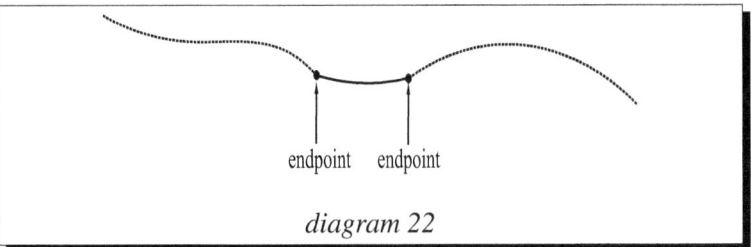

diagram 22

(φ_6^*) $\quad \varphi_6^*(\alpha_1, \alpha_2, ...) \Leftrightarrow_{def} \bigvee_{j<\omega} (\bigwedge_{k<\omega} \alpha_k < \alpha_j)$

(φ_7^*) $\quad \varphi_7^*(\alpha_1, \alpha_2, ...) \Leftrightarrow_{def} \bigvee_{j<\omega} (\bigwedge_{k<\omega} \alpha_j < \alpha_k)$

Now, by putting $\varphi_1 - \varphi_5$ and $\varphi_6^* - \varphi_7^*$ and $\varphi_8 - \varphi_{10}$ together, we get the following definition of a segment $\varsigma(\alpha_1, \alpha_2, ...)$:

$$\varsigma(\alpha_1, \alpha_2, ...) \Leftrightarrow_{def} \begin{array}{l} \bigwedge_{j \leq 5} \varphi_j(\alpha_1, \alpha_2, ...) \wedge \varphi_6^*(\alpha_1, \alpha_2, ...) \wedge \\ \wedge \varphi_7^*(\alpha_1, \alpha_2, ...) \wedge \bigwedge_{8 \leq j \leq 10} \varphi_j(\alpha_1, \alpha_2, ...) \end{array}$$

In order to define a closed line, we need a trickier device. First, it is necessary to note that we always obtain an infinite line if we leave out a point of it (see *diagram 23*).

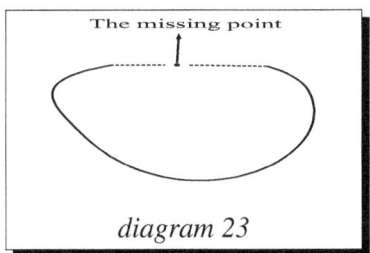

diagram 23

So, we define, first, the set of all the lines from which just one point of each of them is dropped out:

$$\pi(\alpha_1, \alpha_2, ...) \Leftrightarrow_{def} \chi(\alpha_2, \alpha_3, ...) \wedge \chi(\alpha_1, \alpha_3, ...) \wedge ... \\ ... \wedge \chi(..., \alpha_{j-1}, \alpha_{j+1}, ...) \wedge ... \wedge \neg \chi(\alpha_1, \alpha_2, ...)$$

and then, using this definition, we define $\tau(\beta_1, \beta_2, \alpha_1, \alpha_2, ...)$ as a closed line:

$$\tau(\beta_1, \beta_2, \alpha_1, \alpha_2, ...) \Leftrightarrow_{def} \begin{array}{l} \varsigma(\alpha_1, \alpha_2, ...) \wedge \bigvee_{i<\omega} \beta_1 = \alpha_i \wedge \bigvee_{i<\omega} \beta_2 = \alpha_i \wedge \\ \wedge \bigwedge_{j<\omega} \alpha_j < \beta_1 \wedge \bigwedge_{j<\omega} \beta_2 < \alpha_j \end{array}$$

Then, finally, we define the *full two-dimensional circle* (see *diagram 24*) as:

$$\mu(\alpha_1, \alpha_2, \ldots) \Leftrightarrow_{def} \begin{array}{l} (\exists \beta_i)_{i<\omega} (\exists \gamma_j)_{j<\omega} (\varsigma(\beta_1, \beta_2, \ldots) \wedge \pi(\gamma_1, \gamma_2, \ldots) \wedge \\ \wedge \bigvee_{k,l<\omega} \tau(\gamma_k, \gamma_l, \beta_1, \beta_2, \ldots) \wedge \\ \wedge (\bigvee_{m,n<\omega} \alpha_m = \beta_n \vee \bigvee_{m,n<\omega} \alpha_m = \gamma_n) \end{array}$$

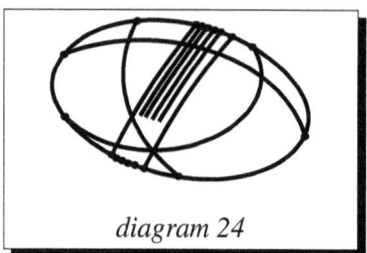

diagram 24

In order to secure that the identity of two regions in S_R when spoken of in S_P is just about the unique "region" we need a function f^*, mapping the variables of S_R into an infinite-tuple of variables of S_P:

$$f_1^* : a_n \to (\alpha_{n_1}, \alpha_{n_2}, \ldots) \qquad (n=1,2,\ldots)$$

Now, in view of all this, the two regions a_m and a_n identical in S_R should be spoken of in S_P in accordance with the following translation rule:

(T_1^*) $\qquad a_m = a_n =^{T_1^*}$
$\qquad (\mu(\alpha_{m_1}, \alpha_{m_2}, \ldots) \wedge \mu(\alpha_{n_1}, \alpha_{n_2}, \ldots)) \to \bigwedge_{i<\omega} \alpha_{m_i} = \alpha_{n_i}$

In order to secure that by speaking about a line in S_R and in S_P we are speaking of one and the same line, we need the function f_2^* mapping an infinite-tuple of variables of S_R into an infinite-tuple of variables of S_P:

$$f_2^* : (a_1, a_2, \ldots) \to (\alpha_1, \alpha_2, \ldots)$$

The next translation rule concerns speaking of the line within S_P into the speaking of the corresponding line of S_R. It would be just the inverse of the translation rule (T_2) above:

(T_2^*) $\qquad a_1 <^* a_2 \wedge a_2 <^* a_3 \wedge \ldots =^{T_2^*} \chi(\alpha_1, \alpha_2, \ldots)$

And now, finally, given that all the relations in S_R are defined via |, the only remaining translation rule that we need is the rule concerning the connection of two regions. So, given the function f_1^*, the translation rule (T_3^*) reads as follows:

(T_3^*) $\quad a_m \mid a_n =^{T_3^*} \mu(\alpha_{m_1}, \alpha_{m_2}, ...) \wedge \mu(\alpha_{n_1}, \alpha_{n_2}, ...) \wedge \bigvee_{i,j<\omega} \alpha_{m_i} = \alpha_{n_j}$

5. Proof that S_R and S_P are only trivially different in the generalized sense

5.1. Proof that all the axioms of S_R are theorems of S_P

Translation of $(AS_R 1)$
According to $(T_1^*) - (T_3^*)$, the axiom

$(AS_R 1)$ $\quad \forall a \, (a \mid a)$

reads in S_P as follows:

$(\forall \alpha_n)_{n<\omega} (\mu(\alpha_1, \alpha_2, ...) \rightarrow (\mu(\alpha_1, \alpha_2, ...) \wedge \mu(\alpha_1, \alpha_2, ...) \wedge \bigvee_{i,j<\omega} \alpha_i = \alpha_j))$

that is, after the double application of the contraction rule:

$(\forall \alpha_n)_{n<\omega} (\mu(\alpha_1, \alpha_2, ...) \rightarrow \bigvee_{i,j<\omega} \alpha_i = \alpha_j)$

The last formula is trivially true in S_P, since every two dimensional circle indeed has at least one point in common with one point from the set of points from which it consists.

Translation of $(AS_R 2)$
According to $(T_1^*) - (T_3^*)$ the axiom

$(AS_R 2)$ $\quad \forall a \forall b \, (a \mid b \rightarrow b \mid a),$

reads in S_P as follows:
$(\forall \alpha_{m_i})_{i<\omega} (\forall \alpha_{n_j})_{j<\omega} ((\mu(\alpha_{m_1}, \alpha_{m_2}, ...) \wedge \mu(\alpha_{n_1}, \alpha_{n_2}, ...) \wedge$
$\wedge \bigvee_{i,j<\omega} \alpha_{m_i} = \alpha_{n_j}) \rightarrow (\mu(\alpha_{n_1}, \alpha_{n_2}, ...) \wedge \mu(\alpha_{m_1}, \alpha_{m_2}, ...) \wedge$
$\wedge \bigvee_{i,j<\omega} \alpha_{n_i} = \alpha_{m_j}))$

which gives:
$(\forall \alpha_{m_i})_{i<\omega} (\forall \alpha_{n_j})_{j<\omega} ((\mu(\alpha_{m_1}, \alpha_{m_2}, ...) \wedge \mu(\alpha_{n_1}, \alpha_{n_2}, ...) \wedge$
$\wedge \bigvee_{i,j<\omega} \alpha_{m_i} = \alpha_{n_j}) \rightarrow \bigvee_{i,j<\omega} \alpha_{n_i} = \alpha_{m_j})$

The last formula is also trivially true in S_P, since two identical, overlapping or touching two-dimensional circles have at least one common point.

Translation of $(AS_R 3)$
According to $(T_1^*) - (T_3^*)$ the axiom

$(AS_R 3)$ $\quad \forall a \forall b \, (\forall c \, (c \mid a \leftrightarrow c \mid b) \rightarrow a = b).$

reads in S_P as follows:

$$(\forall \alpha_{m_i})_{i<\omega}(\forall \alpha_{n_j})_{j<\omega}(\mu(\alpha_{m_1},\alpha_{m_2},\ldots) \wedge \mu(\alpha_{n_1},\alpha_{n_2},\ldots) \rightarrow$$
$$\rightarrow (\forall \alpha_{k_l})_{l<\omega}(\mu(\alpha_{k_1},\alpha_{k_2},\ldots) \rightarrow$$
$$\rightarrow (\mu(\alpha_{m_1},\alpha_{m_2},\ldots) \wedge \mu(\alpha_{k_1},\alpha_{k_2},\ldots) \wedge \bigvee_{i,l<\omega} \alpha_{m_i} = \alpha_{k_l}) \leftrightarrow$$
$$\leftrightarrow (\mu(\alpha_{m_1},\alpha_{m_2},\ldots) \wedge \mu(\alpha_{k_1},\alpha_{k_2},\ldots) \wedge \bigvee_{i,l<\omega} \alpha_{m_i} = \alpha_{k_l}) \rightarrow$$
$$\rightarrow (\mu(\alpha_{m_1},\alpha_{m_2},\ldots) \wedge \mu(\alpha_{n_1},\alpha_{n_2},\ldots) \rightarrow \bigwedge_{i<\omega} \alpha_{m_i} = \alpha_{n_i}))$$

which gives:

$$(\forall \alpha_{m_i})_{i<\omega}(\forall \alpha_{n_j})_{j<\omega}(\mu(\alpha_{m_1},\alpha_{m_2},\ldots) \wedge \mu(\alpha_{n_1},\alpha_{n_2},\ldots) \rightarrow (\forall \alpha_{k_l})_{l<\omega}$$
$$(\mu(\alpha_{k_1},\alpha_{k_2},\ldots) \rightarrow (\bigvee \alpha_{m_i} = \alpha_{k_l} \leftrightarrow \bigvee_{i,l<\omega} \alpha_{m_i} = \alpha_{k_l}) \rightarrow$$
$$\rightarrow \bigwedge_{i<\omega} \alpha_{m_i} = \alpha_{n_i}$$

The last formula is obviously true in S_P, for if it holds for two two-dimensional circles that every two-dimensional circle that has at least one point common to one of them has also at least one point common with the other one, the two two-dimensional circles must be identical to each other.

Translation of (AS_R4)

In order to see what we get, by translating the axiom

(AS_R4) $\forall a \forall b \forall c((b \triangleleft a \wedge c \triangleleft a) \rightarrow \exists d((d \mid c \wedge d \mid b) \wedge \forall e(e \triangleleft d \rightarrow e \triangleleft a)))$

into S_P, let us first see what the relations $\sqsubseteq$, $\sqsubset$, $\circ$, ∞ and $\triangleleft$ of the system S_R look like in S_P:

- $(\alpha_{m_1},\alpha_{m_2},\ldots) \sqsubseteq_p (\alpha_{n_1},\alpha_{n_2},\ldots) \Leftrightarrow_{def} \mu(\alpha_{m_1},\alpha_{m_2},\ldots) \wedge \mu(\alpha_{n_1},\alpha_{n_2},\ldots) \wedge$
  $\wedge (\forall \alpha_{k_l})_{l<\omega}(\mu(\alpha_{k_1},\alpha_{k_2},\ldots) \rightarrow (\bigvee \alpha_{m_i} = \alpha_{k_l} \rightarrow \bigvee_{i,l<\omega} \alpha_{m_i} = \alpha_{k_l}))$

- $(\alpha_{m_1},\alpha_{m_2},\ldots) \sqsubset_p (\alpha_{n_1},\alpha_{n_2},\ldots) \Leftrightarrow_{def} (\alpha_{m_1},\alpha_{m_2},\ldots) \sqsubseteq_p (\alpha_{n_1},\alpha_{n_2},\ldots) \wedge$
  $\wedge \neg \bigwedge_{i<\omega} \alpha_{m_i} = \alpha_{n_i}$

- $(\alpha_{m_1},\alpha_{m_2},\ldots) \circ_p (\alpha_{n_1},\alpha_{n_2},\ldots) \Leftrightarrow_{def}$
  $\Leftrightarrow_{def} (\exists \alpha_{k_l})_{l<\omega}(\mu(\alpha_{k_1},\alpha_{k_2},\ldots) \wedge (\alpha_{k_1},\alpha_{k_2},\ldots) \sqsubseteq_p$
  $(\alpha_{m_1},\alpha_{m_2},\ldots) \wedge (\alpha_{k_1},\alpha_{k_2},\ldots) \sqsubseteq_p (\alpha_{n_1},\alpha_{n_2},\ldots))$

- $(\alpha_{m_1},\alpha_{m_2},\ldots) \infty_p (\alpha_{n_1},\alpha_{n_2},\ldots) \Leftrightarrow_{def} \mu(\alpha_{m_1},\alpha_{m_2},\ldots) \wedge \mu(\alpha_{n_1},\alpha_{n_2},\ldots) \wedge$
  $\wedge \bigvee_{i,j<\omega} \alpha_{m_i} = \alpha_{n_j} \wedge \neg (\alpha_{m_1},\alpha_{m_2},\ldots) \circ_p (\alpha_{n_1},\alpha_{n_2},\ldots)$

- $(\alpha_{m_1},\alpha_{m_2},\ldots) \triangleleft_p (\alpha_{n_1},\alpha_{n_2},\ldots) \Leftrightarrow_{def}$
  $\Leftrightarrow_{def} (\alpha_{m_1},\alpha_{m_2},\ldots) \sqsubset_p (\alpha_{n_1},\alpha_{n_2},\ldots) \wedge$
  $(\forall \alpha_{k_l})_{l<\omega}((\alpha_{k_1},\alpha_{k_2},\ldots) \infty_p (\alpha_{m_1},\alpha_{m_2},\ldots) \rightarrow$
  $\rightarrow \neg (\alpha_{k_1},\alpha_{k_2},\ldots) \infty_p (\alpha_{n_1},\alpha_{n_2},\ldots))$

In view of this, the axiom (AS_R4) reads in S_P as follows:

Gunkology and Pointilism

$$(\forall \alpha_{m_i})_{i<\omega} (\forall \alpha_{n_j})_{j<\omega} (\forall \alpha_{k_l})_{l<\omega} ((\alpha_{n_1}, \alpha_{n_2}, ...) \triangleleft_p (\alpha_{m_1}, \alpha_{m_2}, ...) \wedge$$
$$\wedge (\alpha_{k_1}, \alpha_{k_2}, ...) \triangleleft_p (\alpha_{m_1}, \alpha_{m_2}, ...) \rightarrow$$
$$\rightarrow (\exists \alpha_{r_e})_{e<\omega} (\mu(\alpha_{r_1}, \alpha_{r_2}, ...) \wedge \bigvee_{l,e<\omega} \alpha_{r_e} = \alpha_{k_l} \wedge$$
$$\wedge \bigvee_{j,e<\omega} \alpha_{r_e} = \alpha_{n_j} \wedge (\forall \alpha_{s_f})_{f<\omega} ((\alpha_{s_1}, \alpha_{s_2}, ...) \triangleleft_p (\alpha_{r_1}, \alpha_{r_2}, ...) \rightarrow$$
$$\rightarrow (\alpha_{s_1}, \alpha_{s_2}, ...) \triangleleft_p (\alpha_{m_1}, \alpha_{m_2}, ...)))$$

The last formula is true in S_P because, according to the definition of μ, what is required is that any two end-points of the region that consists of two separate two-dimensional circles can be connected by a line segment that contains only the points from the set of points of which the two two-dimensional circles consists, which is impossible.

Translation of (AS_R5)
According to $(T_1^*) - (T_3^*)$ and the way in which $\triangleleft_p$ is to be understood in S_P, the axiom

(AS_R5) $\forall a \exists b \, (b \triangleleft a)$

reads as follows:

$$(\forall \alpha_{m_i})_{i<\omega}(\mu(\alpha_{m_1}, \alpha_{m_2}, ...) \rightarrow$$
$$\rightarrow (\exists \alpha_{n_j})_{j<\omega}(\mu(\alpha_{n_1}, \alpha_{n_2}, ...) \wedge ((\alpha_{n_1}, \alpha_{n_2}, ...) \triangleleft_p (\alpha_{m_1}, \alpha_{m_2}, ...))))$$

This formula is trivially true in S_P, since it holds for every two-dimensional circle that there is a two-dimensional circle that consists of its points only but does not contain all of them.

Translation of (AS_R6)
According to $(T_1^*) - (T_3^*)$ and the way in which $\triangleleft_p$ is to be understood in S_P, the axiom

(AS_R6) $\forall a \exists b \, (a \triangleleft b)$

reads as follows:

$$(\forall \alpha_{m_i})_{i<\omega}(\mu(\alpha_{m_1}, \alpha_{m_2}, ...) \rightarrow$$
$$\rightarrow (\exists \alpha_{n_j})_{j<\omega}(\mu(\alpha_{n_1}, \alpha_{n_2}, ...) \wedge ((\alpha_{m_1}, \alpha_{m_2}, ...) \triangleleft_p (\alpha_{n_1}, \alpha_{n_2}, ...))))$$

The last formula is true in S_P, since no two-dimensional circle covers the whole infinite two-dimensional surface, so that it holds for every two-dimensional circle that there is a two-dimensional circle that contains all but also more points than it.

Translation of (AS_R7)
According to $(T_1^*) - (T_3^*)$ and the way in which $\triangleleft_p$ and $\sqsubseteq_p$ are to be understood in S_P, the axiom

(AS_R7) $\forall a \forall b (\forall c (b \triangleleft c \to c \circ a) \to b \triangleleft a)$

reads as follows

$$(\forall \alpha_{m_i})_{i<\omega} (\forall \alpha_{n_j})_{j<\omega} ((\forall \alpha_{k_l})_{l<\omega} ((\alpha_{n_1}, \alpha_{n_2}, \ldots) \triangleleft_p (\alpha_{k_1}, \alpha_{k_2}, \ldots) \to$$
$$\to (\alpha_{k_1}, \alpha_{k_2}, \ldots) \circ_p (\alpha_{m_1}, \alpha_{m_2}, \ldots)) \to (\alpha_{n_1}, \alpha_{n_2}, \ldots) \triangleleft_p (\alpha_{m_1}, \alpha_{m_2}, \ldots))$$

Given the way in which the regions are translated as sets of points of S_P, the last formula says nothing else but that there is no two-dimensional circle within which there could be a set of points that would not be a set of points of the two-dimensional circle itself.

Translation of (AS_R8)

In order to see what we get, by translating the axiom

(AS_R8) $\forall a, b, c_1, d_1, c_2, d_2 (W(a,b,c_1,d_1) \land W(a,b,c_2,d_2) \to$
$\to D(c_1, c_2, d_1) \land D(c_1, c_2, d_2))$

into S_P, let us first see what the relations $\ll$, B, N, W, P and D of the system S_R look like in S_P:

- $(\alpha_{m_1}, \alpha_{m_2}, \ldots) \ll_p (\alpha_{n_1}, \alpha_{n_2}, \ldots) \Leftrightarrow_{def} ((\alpha_{m_1}, \alpha_{m_2}, \ldots) \sqsubset_p (\alpha_{n_1}, \alpha_{n_2}, \ldots) \land$
  $\land (\forall \alpha_{k_l})_{l<\omega} ((\alpha_{m_1}, \alpha_{m_2}, \ldots) \triangleleft_p (\alpha_{k_1}, \alpha_{k_2}, \ldots) \to$
  $\to \neg (\alpha_{k_1}, \alpha_{k_2}, \ldots) \sqsubseteq_p (\alpha_{n_1}, \alpha_{n_2}, \ldots))$

- $B_p (\alpha_{m_i}, \alpha_{n_j}, \alpha_{k_l})_{i,j,l<\omega} \Leftrightarrow_{def} (\alpha_{n_1}, \alpha_{n_2}, \ldots) \ll_p (\alpha_{m_1}, \alpha_{m_2}, \ldots) \land$
  $\land (\alpha_{k_1}, \alpha_{k_2}, \ldots) \ll_p (\alpha_{m_1}, \alpha_{m_2}, \ldots) \land (\alpha_{n_1}, \alpha_{n_2}, \ldots) \infty_p (\alpha_{k_1}, \alpha_{k_2}, \ldots) \land$
  $\land \neg (\exists \alpha_{r_e})_{e<\omega} (((\alpha_{r_1}, \alpha_{r_2}, \ldots) \infty_p (\alpha_{n_1}, \alpha_{n_2}, \ldots) \lor$
  $\lor (\alpha_{r_1}, \alpha_{r_2}, \ldots) \infty_p (\alpha_{k_1}, \alpha_{k_2}, \ldots)) \land$
  $\land (\alpha_{r_1}, \alpha_{r_2}, \ldots) \sqsubset_p (\alpha_{m_1}, \alpha_{m_2}, \ldots) \land$
  $\land \left(\forall \alpha_{s_f} \right)_{f<\omega} ((\alpha_{r_1}, \alpha_{r_2}, \ldots) \sqsubseteq_p (\alpha_{s_1}, \alpha_{s_2}, \ldots) \land$
  $\land (\alpha_{s_1}, \alpha_{s_2}, \ldots) \ll_p (\alpha_{m_1}, \alpha_{m_2}, \ldots) \to$
  $\to (\alpha_{s_1}, \alpha_{s_2}, \ldots) \circ_p (\alpha_{n_1}, \alpha_{n_2}, \ldots) \lor (\alpha_{s_1}, \alpha_{s_2}, \ldots) \circ_p (\alpha_{k_1}, \alpha_{k_2}, \ldots)))$

- $N_p (\alpha_{m_i}, \alpha_{n_j}, \alpha_{k_l}, \alpha_{r_e})_{i,j,l,e<\omega} \Leftrightarrow_{def} (\alpha_{m_1}, \alpha_{m_2}, \ldots) \infty_p (\alpha_{n_1}, \alpha_{n_2}, \ldots) \land$
  $\land (\alpha_{m_1}, \alpha_{m_2}, \ldots) \infty_p (\alpha_{r_1}, \alpha_{r_2}, \ldots) \land (\alpha_{m_1}, \alpha_{m_2}, \ldots) \infty_p (\alpha_{s_1}, \alpha_{s_2}, \ldots) \land$
  $\land (\alpha_{n_1}, \alpha_{n_2}, \ldots) \infty_p (\alpha_{r_1}, \alpha_{r_2}, \ldots) \land (\alpha_{n_1}, \alpha_{n_2}, \ldots) \infty_p (\alpha_{s_1}, \alpha_{s_2}, \ldots) \land$
  $\land (\alpha_{r_1}, \alpha_{r_2}, \ldots) \infty_p (\alpha_{s_1}, \alpha_{s_2}, \ldots)$

- $W_p\left(\alpha_{m_i}, \alpha_{n_j}, \alpha_{k_l}, \alpha_{r_e}\right)_{i,j,l,e<\omega} \Leftrightarrow_{def}$
  $\left(\exists \alpha_{s_f}\right)_{f<\omega} (B_p((\alpha_{s_1}, \alpha_{s_2}, \ldots), (\alpha_{m_1}, \alpha_{m_2}, \ldots), (\alpha_{n_1}, \alpha_{n_2}, \ldots)) \wedge$
  $\wedge N_p((\alpha_{m_1}, \alpha_{m_2}, \ldots), (\alpha_{n_1}, \alpha_{n_2}, \ldots), (\alpha_{k_1}, \alpha_{k_2}, \ldots), (\alpha_{r_1}, \alpha_{r_2}, \ldots)) \wedge$
  $\wedge (\alpha_{k_1}, \alpha_{k_2}, \ldots) \sqsubseteq_p (\alpha_{s_1}, \alpha_{s_2}, \ldots) \wedge (\alpha_{r_1}, \alpha_{r_2}, \ldots) \sqsubseteq_p (\alpha_{s_1}, \alpha_{s_2}, \ldots) \wedge$
  $\wedge (\forall \alpha_{v_o})_{o<\omega}((\alpha_{v_1}, \alpha_{v_2}, \ldots) \sqsubseteq_p (\alpha_{s_1}, \alpha_{s_2}, \ldots) \wedge$
  $\wedge (\mu(\alpha_{v_1}, \alpha_{v_2}, \ldots) \wedge \mu(\alpha_{k_1}, \alpha_{k_2}, \ldots)) \wedge \bigvee_{l,o<\omega} \alpha_{v_o} = \alpha_{k_l} \wedge$
  $\wedge ((\mu(\alpha_{v_1}, \alpha_{v_2}, \ldots) \wedge \mu(\alpha_{r_1}, \alpha_{r_2}, \ldots)) \wedge \bigvee_{s,o<\omega} \alpha_{v_o} = \alpha_{r_s}) \rightarrow$
  $\rightarrow ((\mu(\alpha_{v_1}, \alpha_{v_2}, \ldots) \wedge \mu(\alpha_{m_1}, \alpha_{m_2}, \ldots)) \wedge \bigvee_{i,o<\omega} \alpha_{v_o} = \alpha_{m_i})$

- $(\alpha_{m_1}, \alpha_{m_2}, \ldots) P_p (\alpha_{n_1}, \alpha_{n_2}, \ldots) \Leftrightarrow_{def} (\exists \alpha_{k_l})_{l<\omega} (\exists \alpha_{r_e})_{e<\omega}$
  $W_p((\alpha_{m_1}, \alpha_{m_2}, \ldots), (\alpha_{n_1}, \alpha_{n_2}, \ldots), (\alpha_{k_1}, \alpha_{k_2}, \ldots), (\alpha_{r_1}, \alpha_{r_2}, \ldots))$

- $D_p((\alpha_{m_1}, \alpha_{m_2}, \ldots), (\alpha_{n_1}, \alpha_{n_2}, \ldots), (\alpha_{k_1}, \alpha_{k_2}, \ldots)) \Leftrightarrow_{def}$
  $(\alpha_{m_1}, \alpha_{m_2}, \ldots) \sqsubseteq_p (\alpha_{n_1}, \alpha_{n_2}, \ldots) \vee (\alpha_{n_1}, \alpha_{n_2}, \ldots) \sqsubseteq_p (\alpha_{m_1}, \alpha_{m_2}, \ldots) \vee$
  $\vee (\alpha_{m_1}, \alpha_{m_2}, \ldots) \circ_p (\alpha_{n_1}, \alpha_{n_2}, \ldots) \vee (\alpha_{m_1}, \alpha_{m_2}, \ldots) \sqsubseteq_p (\alpha_{k_1}, \alpha_{k_2}, \ldots) \vee$
  $\vee (\alpha_{k_1}, \alpha_{k_2}, \ldots) \sqsubseteq_p (\alpha_{m_1}, \alpha_{m_2}, \ldots) \vee (\alpha_{m_1}, \alpha_{m_2}, \ldots) \circ_p (\alpha_{k_1}, \alpha_{k_2}, \ldots)$

In view of all this, the axiom (AS_R8) reads in S_P as follows:

$(\forall \alpha_{m_i})_{i<\omega} (\forall \alpha_{n_j})_{j<\omega} (\forall \alpha_{k_l})_{l<\omega} (\forall \alpha_{r_e})_{e<\omega} \left(\forall \alpha_{s_f}\right)_{f<\omega} (\forall \alpha_{v_o})_{o<\omega}$
$(W_p(\alpha_{m_i}, \alpha_{n_j}, \alpha_{k_l}, \alpha_{r_e})_{i,j,l,e<\omega} \wedge W_p\left(\alpha_{m_i}, \alpha_{n_j}, \alpha_{s_f}, \alpha_{v_o}\right)_{i,j,f,o<\omega} \rightarrow$
$\rightarrow D_p((\alpha_{k_1}, \alpha_{k_2}, \ldots), (\alpha_{s_1}, \alpha_{s_2}, \ldots), (\alpha_{r_1}, \alpha_{r_2}, \ldots)) \wedge$
$\wedge D_p((\alpha_{k_1}, \alpha_{k_2}, \ldots), (\alpha_{s_1}, \alpha_{s_2}, \ldots), (\alpha_{v_1}, \alpha_{v_2}, \ldots)))$

In order to see that the last formula must be true in S_P, it is sufficient to realize that, according to (AS_P1), there can be no lines that are skew.

Translation of (AS_R9)
In order to see what we get, by translating the axiom

(AS_R9) $\quad \forall a (\forall b_n)_{n<\omega} (\varphi(a, b_1, b_2, \ldots) \rightarrow \exists c \psi(a, c, b_1, b_2, \ldots))$

into S_P, let us first see how the following two formulae

- $\varphi(a, b_1, b_2, \ldots) \Leftrightarrow_{def}$
  $\Leftrightarrow_{def} \bigwedge_{n<\omega} b_n \triangleleft a \wedge \bigwedge_{m<n<\omega} b_m \triangleleft b_n \wedge \neg \exists c (\bigwedge_{n<\omega} b_n \triangleleft c \wedge c \triangleleft a)$
- $\psi(a, b, c_1, c_2, \ldots) \Leftrightarrow_{def} b \infty a \wedge \forall d (b \triangleleft d \rightarrow \bigvee_{n<\omega} c_n \mid d)$

of the system S_R look like in S_P. Recall that the symbol $\vec{\alpha}$ is used to denote a countable sequence of variables $\alpha_1, \alpha_2, \ldots$, with or without a subscript.

- $\varphi_p\left(\vec{\alpha}, \vec{\beta_1}, \vec{\beta_2}, \ldots\right) \Leftrightarrow_{def}$
  $\Leftrightarrow_{def} \mu(\vec{\alpha}) \wedge \bigwedge_{n<\omega} \mu\left(\vec{\beta_n}\right) \wedge \bigwedge_{n<\omega} \vec{\beta_n} \triangleleft_p \vec{\alpha} \wedge \bigwedge_{m<n<\omega} \vec{\beta_m} \triangleleft \vec{\beta_n} \wedge$
  $\wedge \neg \exists \vec{\gamma} (\mu(\vec{\gamma}) \wedge \bigwedge_{n<\omega} \vec{\beta_n} \triangleleft_p \vec{\gamma} \wedge \vec{\gamma} \triangleleft_p \vec{\alpha})$

- $\psi_p\left(\vec{\alpha}, \vec{\beta}, \vec{\gamma_1}, \vec{\gamma_2}, \ldots\right) \Leftrightarrow_{def}$
  $\Leftrightarrow_{def} \mu(\vec{\alpha}) \wedge \mu(\vec{\beta}) \wedge \bigwedge_{n<\omega} \mu(\vec{\gamma_n}) \wedge \vec{\beta} \infty_p \vec{\alpha} \wedge$
  $\forall \vec{\delta} \left(\mu(\vec{\delta}) \wedge \vec{\beta} \triangleleft_p \vec{\delta} \rightarrow \bigvee_{n<\omega} \vec{\gamma_n} |_p \vec{\delta} \right)$

In view of all this, the axiom (AS_R9) reads in S_P as follows:

$$\forall \vec{\alpha} \left(\forall \vec{\beta_n} \right)_{n<\omega} (\mu(\vec{\alpha}) \wedge \bigwedge_{n<\omega} \mu(\vec{\beta_n}) \wedge \varphi_p\left(\vec{\alpha}, \vec{\beta_1}, \vec{\beta_2}, \ldots\right) \rightarrow$$
$$\rightarrow \exists \vec{\gamma} (\mu(\vec{\gamma}) \wedge \psi_p(\vec{\alpha}, \vec{\gamma}, \vec{\beta_1}, \vec{\beta_2}, \ldots)))$$

It is clear that the last formula is true in S_P. since the continuity of the parallels mentioned precludes the existence of a "gap" in any directions.

5.2. Proof that the axiom of S_P is a theorem of S_R

In order to see what we get by translating the axiom

(AS_P1) $\quad (\exists \vec{\alpha}^n)_{n<\omega} (\exists \vec{\beta}^m)_{m<\omega} (\psi(\vec{\alpha}^1, \vec{\alpha}^2, \ldots) \wedge \psi(\vec{\beta}^1, \vec{\beta}^2, \ldots) \wedge$
$\quad \wedge \alpha_j^i = \beta_k^l \wedge \neg \alpha_j^i = \beta_q^p \wedge \forall \gamma (\bigvee_{m,n<\omega} \gamma = \alpha_n^m)),$
$\quad$ for $l \neq p, k \neq q$ and $i, j, l, k, p, q < \omega$

into S_R, let us first see what the following formula
$\psi(\vec{\alpha}^1, \vec{\alpha}^2, \ldots) \Leftrightarrow_{def} \bigwedge_{j \leq 10} \bigwedge_{i<\omega} \varphi_j(\vec{\alpha}^i) \wedge \neg \exists x, y (\alpha_n^i = x = y = \alpha_m^j),$
for $i, j, m, n < \omega$ and $i \neq j$

of the system S_P looks like in S_R:

$\psi_r(\vec{a}^1, \vec{a}^2, \ldots) \Leftrightarrow_{def} a_1^1 <^* a_2^1 \wedge a_2^1 <^* a_3^1 \wedge \ldots \wedge$
$\wedge a_1^2 <^* a_2^2 \wedge a_2^2 <^* a_3^2 \wedge \ldots \wedge \neg \exists x, y (a_n^i = x = y = a_m^j),$
for $i, j, m, n < \omega$ and $i \neq j$

To see what this means, let us first remember that the supposed function f_2 tied to the translation rule T_2 implies that we always speak about a particular line selected by this function so that it holds that according to T_2 the translation rule $\chi(\alpha_1, \alpha_2, \ldots) =^{T_2} a_1 <^* a_2 \wedge a_2 <^* a_3 \wedge \ldots$ says that in S_R we speak only of the regions represented by *diagram 21* (p. 151), that is, of the regions constituting that which is a line in S_R. The formula $\psi_r(\vec{a}^1, \vec{a}^2, \ldots)$ speaks of lines being "parallel" by asserting that no two regions from the defining set of regions making up a line can be identical, for otherwise the lines defined by them would share a common part, however small.

In view of all this, the axiom $(AS_P 1)$ reads in S_R as follows:

$$(\exists \vec{a}^n)_{n<\omega}(\exists \vec{b}^m)_{m<\omega}(\psi_r(\vec{a}^1,\vec{a}^2,...) \wedge \psi_r(\vec{b}^1,\vec{b}^2,...) \wedge$$
$$\wedge a_j^i \mid b_k^l \wedge \neg a_j^i \mid b_q^p \wedge \forall c(\bigvee_{m,n<\omega} c \mid a_n^m)),$$
$$\text{for } l \neq p, k \neq q \text{ and } i,j,l,k,p,q < \omega$$

The crucial but also a rather tricky thing is to see exactly why all the axioms of S_R are needed to secure the truth of the translation of $(AS_P 1)$. Quite generally, the connectedness axioms $(AS_R 1)$ - $(AS_R 3)$ are needed because they express primitive properties of regions without which it would obviously be impossible to speak of points and lines. In particular, the axiom $(AS_R 4)$ precludes the possibility that two disconnected "parts" of one region make up one and the same line; the axioms $(AS_R 5)$ - $(AS_R 6)$ guarantee the existence of infinitely many regions needed for the very formulation of the axiom $(AS_P 1)$; the axiom $(AS_R 7)$, which precludes the existence of doughnut-like regions, is needed to secure the connectedness of every region to some region of which the translation of $(AS_P 1)$ speaks; the axiom $(AS_R 8)$ is necessary for precluding the possibility that the lines the axiom $(AS_P 1)$ is speaking about are skew; and finally, the axiom $(AS_R 9)$ secures the continuity of regions necessary for the continuity of lines.

6. Metalogical and Metaontological Consequences

In view of the above results, two general conclusions, one metalogical and one metaontological, can be straightforwardly derived.

From the metalogical point of view, it follows that the two formal theories, the Point-Based and the Region-Based theory, have the same truth-expressive power in relation to the representation of the infinite two-dimensional continuum. So, it hardly makes any sense to raise, at least *ceteris paribus*, the metalogical question of whether it is the Point-Based rather than the Region-Based or the Region-Based rather the Point-Based theory that which represents *the true theory* of the infinite two-dimensional continuum. Given that all truths and only truths of any of the two theories can be expressed as truths of any of the theories, it follows that if any of the two theories is true, the other one is true as well.

From the metaontological point of view, it follows that the infinite two-dimensional continuum can be analyzed by starting with a set of null-dimensional points, with one-dimensional lines and two-dimensional regions supervening on them, as well as by starting with a set of two-dimensional regions, with one-dimensional lines and null-dimensional points supervening on them. So again, it hardly makes any sense to raise, at least *ceteris paribus*, the metaontological question of whether the points rather than regions or the regions rather than points represent *the basic elements* of the infinite two-dimensional continuum. Given that regions can be said to supervene on points just as it can be said that points supervene on regions, there can be *no ontological priority* in view of these two kinds of entities.

Moreover, given the translation rules (T_2) and (T_2^*), according to which lines are to be spoken of in S_P and S_R respectively, along with the way in which it is shown by Arsenijević and Kapetanović[12] how points can be spoken of in the Interval-Based System S_I and the way in which two sets of parallels are used above for speaking of the regions of the infinite plain, we can even generalize the story and conclude that, metalogically, the Point-Based System S_P, the Interval-Based System S_I and the Region-Based System S_R, as formal theories of the two-dimensional continuum, are all equivalent in view of their truth-expressive power, while, metaontologically, there can be no ontological priority within the infinite two-dimensional continuum between null-dimensional points, one-dimensional intervals and two-dimensional regions.

In view of all this, it is curious why the Aristotelian and the Cantorian theory of the continuum (or of the one-dimensional and two-dimensional continua at least) have still being considered as two rival theories, while the struggle between Gunkologists and Pointillists has been always cited as an instance of the ontologically important disagreement. It is true, from a historical point of view, that original Aristotelian theory turned out to be insufficient in view of the fact that it lacked the second Cantor condition for the one-dimensional continuity[13], which is expressed above through φ_9 and φ_{10} which are defined in Section **3** above. But since this condition has become known, it should have been no problem for Neo-Aristotelians to adjust it in order to complete the Aristotelian Interval-Based System of the linear continuum[14] as well as to complete the Aristotelian Region-Based System of the two-dimensional continuum (as it is done above by the introduction of *the Continuity Axiom* $(AS_r 9)$). It must be that some deep-rooted prejudice has prevented philosophers to find out a peace-making strategy that would en-

---

[12] See Arsenijević and Kapetanović (2008a) and Arsenijević and Kapetanović (2008b).
[13] See Cantor (1962), pp. 194-195.
[14] As it is done in Arsenijević and Kapetanović (2008a) and Arsenijević and Kapetanović (2008b).

able us to overcome the great struggle between two parties. In what follows, we shall try to give a diagnosis of the phenomenon.

6.1. Rejection of Quine's Semantico-Ontological Slogan

Due to the fact that Aristotle accepted that which he called *Zeno's Axiom*[15], which says that no entity of a higher dimension can consist of entities of a lower dimension, he considered the three-dimensional bodies as basic entities, with surfaces, lines and points as limits supervening on them. Analogously, he considered periods as basic elements of time, with instants only supervening on them. One can understand that after Cantor's introduction of the second continuity condition, which makes it possible to say that the entities of higher dimensions consist of null-dimensional entities, both mathematicians and philosophers have become prone to accept his analysis of the continuum as the right one, since null-dimensional entities look as the best candidates to be considered as proper elements of the continuum, given that only they do not contain either further null-dimensional entities or any other entities as their parts.[16] But this cannot be the whole explanation of the fact that no deeper comparison between the two theories of the continuum has been done.

When in the last three decades of the last century a considerable number of philosophers tried to revive the Aristotelian approach, both within philosophy of time as well as within philosophy of space, they did it, at least tacitly, as if the Aristotelian theory should be reconsidered as an alternative to Cantor's theory, and not in order to investigate the question of their possible equivalence.[17] True, van Benthem proclaimed that "systematic connections between point structures and period structures enable to use both perspectives at will".[18] But he didn't offer any new logico-ontological framework within which these "systematic connections" are to be understood and, in particular, he didn't raise the direct question of whether there is a clearly

---
[15] See Aristotle (1831), *Met.* 1001 b 7.
[16] See, for instance, Russell (1903), Ch. XXXV, Russell (1914), Ch. V, Carnap (1928), 1.4, Grünbaum (1952), Grünbaum (1974), Ch. 6, Salmon (1975), Ch. 1, Robinson (1989), Lewis (1994) and Earman and Roberts (2006).
[17] See, for instance, Hamblin (1969), Hamblin (1971), Humberstone (1979), Foldes (1980), Needham (1981), Burgess (1982), Comer (1985), White (1988), Bochman (1990a), Bochman (1990b), van Benthem (1991), Ch. I.3., and van Benthem (1995). Particularly important are the articles of Roeper's (1997) and (2006), given that they concern more-dimensional Aristotelian continua. See also Mormann (2000) and Sambin (2003).
[18] van Benthem (1991), p. 84.

definable sense in which the two theories could be said to be equivalent.

There are two related reasons why the question of the possible equivalence between the two theories has not been further investigated. The first reason has directly to do with Quine's famous semantico-ontological slogan, which says that "to be assumed as an entity is to be reckoned as the value of a variable".[19] The second reason has to do with the fact that the conception of supervenience was, and still has been, applied nearly exclusively within philosophy of mind, when the mental is said to supervene on the physical but not vice versa, so that the cases of the mutual supervenience, where two sets of entities mutually supervene on each other, has never come into the focus of consideration. Let us explain!

If two theories are such that their variables can range in no model over the elements of one and the same basic set, they are, according to *Quine's Slogan*, about two hopelessly different ontologies, because what exists according to one of them does not exist according to the other, and vice versa. The same consequence is present in Model Theory.[20] Quine's Slogan represents the basis for standard differentiation between formal theories that are only notationally different amongst themselves and those which are not. So, according to standard view, the Aristotelian and the Cantorian theory of the continuum represent real alternatives and not just trivially different theories.

As for the reason concerning supervenience, it may seem that the supervenience relation cannot be symmetric. Some set of entities supervenes on some other set of entities just because the latter is the supervenience base on which the former supervenes, and not vice versa. So, even if the theory of supervenience allows us to say that more-dimensional entities also exist somehow in the Cantorian model, they exist in a sense which is different from the sense in which null-dimensional entities exist. If we wanted to accommodate Quine's Slogan to fit in the theory of existence that is involved in the theory of supervenience, we could say that Quine's Slogan concerns that which exists irreducibly, on which everything else that exists, in a secondary sense, is theoretically and ontologically reducible. But even then, the Aristotelian and the Cantorian theory would be, from the very beginning, logically and ontologically non-trivially different, since the supervenience bases of the two theories are radically different.

However, let us consider the following analogy. Suppose that we compare two formalizations of Propositional Calculus, for instance the Hilbert-

---

[19] Quine (1961), p. 13.
[20] See Hodges (1993), pp. 1-2.

Ackermann[21] and the Meredith[22] formalization, whose axioms are different. Now, though the two formalizations do not differ in view of all the truths they express, one could suggest that, in fact, they represent two non-equivalent theories, because they differ in view of which truths are basic and which are only derived. Of course, we would reject such a suggestion as silly and claim that the difference between basic and derived truths is irrelevant in the given case. Consequently, we would continue to hold that the difference in formalization does not mean that the Hilbert-Ackermann and the Meredith formalization represent different theories of Propositional Calculus. But why the difference in view of basic truths doesn't matter in the given case, while the difference in view of basic entities should matter in the case of Gunkology and Pointillism, given that the relation between basic and derived truths is asymmetric just as it is the relation between basic and supervening entities?

There is hardly any essential difference between the two cases, or at least we don't see any. The only reason for remaining stubborn and claiming that there is still an essential difference between the two has only to do with the prejudice condensed in Quine's Semantico-Ontological Slogan, and that is the reason why it ought to be rejected. For our purposes, instead of Quine's Slogan, we should rather accept the Slogan cited more times above: *What matters by comparing two possibly equivalent theories is not the isomorphism or non-isomorphism of the basic sets of their models but the equivalence or non-equivalence of their truth-expressive powers.* This is the essence of Arsenijević's generalized definition of the syntactically and semantically trivial difference between formal theories, which was used above in the proof that the Point-Based and the Region-Based Theory of the two-dimensional continuum, and two ontologies, Pointillism and Gunkology, as two respective mutually supervening models of them, are only trivially different amongst themselves.

6.2. Refutation of the Argument of Arntzenius' and Hawthorne's

Generally, the authors who tried to re-establish the Aristotelian theory of the continuum, several of them cited above,[23] assumed from the very beginning that the Aristotelian and the Cantorian theory are more than trivially different. However, Arntzenius and Hawthorne offered an *argument*[24] that it must be so, which is prima facie convincing and which therefore should be

---
[21] See Hilbert and Ackermann (1968), p. 27.
[22] See Meredith (1954).
[23] See n. 17.
[24] See Artzenius and Hawthorne (2005).

analyzed and refuted.

The argument is based on what Arntzenius and Hawthorne call the *No-Zero Assumption*,[25] to which Gunkologists are allegedly committed. Namely, since "a thing is gunky just in case every part of that thing has proper parts", and since points do not have proper parts, it seems that the gunkologist conception of the continuum implies that there can be no difference between open (or half-open) and closed intervals or between open and closed regions. And then, since the pointillists can speak of such a difference, it follows that there is a real and irreducible difference between Gunkology and Pointillism.

The argument is wrong, because though it is true that regions in the Region-Based Theory are *originally* neither open nor closed, we can yet speak in S_R (as we did above) of that which in S_P is the difference between open and closed regions. For instance, an infinite set of regions of S_R that exhausts completely a given region (cf., for instance, the way in which the axiom $(AS_R 9)$ is introduced above) is exactly that which is an open region in S_P, while, *in contrast* to such an infinite set of regions, the single region itself that is exhausted by the given set of regions, though originally neither open nor closed, turns out to be exactly that which is a closed region in S_P.

It is important to note that the difference between closed and open regions in S_R depends on the presence of *supervening* entities in its model. But *the same* is true of S_P, because the regions *supervene* on points as on basic elements of its model. So, the difference between closed and open regions is by no means more basic in S_P than it is in S_R.

Probably, the strongest intuitive appeal of Arntzenius' and Hawthorne's *No-Zero Assumption* consists in the fact that in S_P we can get a topologically non-homeomorphic figure by dropping out one single point from a given figure. For instance, if we drop out one single point from a set of points that makes up a circle, we get an open line, which is topologically non-homeomorphic with the given circle. This seems impossible to do in S_R. But, unexpectedly, it is exactly that which we managed to do above by using the "tricky device", when we were defining a circle within S_R via the set of the lines such that just one point of each of them is dropped out (see *diagram 23* and further on).

The wrongness of the argument of Arntzenius' and Hawthorne's is based on the false belief that due to the fact that points are not *elements* or *parts* of gunky continua, they are simply non-entities in the gunkologist conception of the continuum. That's why Arntzenius thinks that Aristotle held that

---

[25] Ibid. p. 443.

"there are no instants in time"[26] as well as no points in space,[27] which is straightforwardly wrong. When speaking of *Zeno's Axiom*, Aristotle did accept that entities of a higher dimension do not consist of entities of a lower dimension,[28] but he explicitly rejected that the latter are *nothing*.[29] It was perhaps Zeno[30] who, by arguing against points as constituents of the magnitude, wanted to pass from "Nothing is added" (because nothing has changed in view of the increasement of a given segment or region) to "That which is added is nothing"[31], but Aristotle rejected this second reading of τὸ προσγινόμενον οὐδέν ἐστιν[32] and accepted that instants, points, lines and surfaces, though not entities in the primary sense (πρῶτον) still exist in a secondary sense.[33] They exist as limits, or, in modern terminology, they supervene on the entities whose limits they are. But independently of Zeno and Aristotle, if we got rid of entities supervening on regions in the models of S_R, we could not speak of lines and regions in the models of S_P as well, because in S_P we have only *points* as the elements of the basic set and not *sets* of points. The power set of the elements of the domain of S_P as well as its elements are not the elements of the domain itself. That's why, in order to see clearly that there is a complete symmetry between S_P and S_R in view of the basic and supervening entities, it is so important to formulate the axioms of S_P and S_R in a pure Hilbertian manner, by letting variables range over individuals only, as we did above.

So, all in all, we should accept that Gunkology and Pointillism are mutually supervening models of the Region-Based and the Point-Based Theory as two only trivially different theories of the two-dimensional continuum.

---

[26] See Arntzenius (2000), p. 187. and p. 202.
[27] See See Arntzenius (2012), Ch. 4.
[28] See Aristotle (1831), *Met.* 1001 b 7.
[29] Ibid. 1001 b 7 ff.
[30] *DK* B **2.**
[31] See Fränkel (1942), p. 199 ff.
[32] *DK* B **2** 13.
[33] For more about this, see Arsenijević, Šćepanović and Massey (2008), pp. 23 ff.

REFERENCES

Aristotle (1831). *Aristotelis opera*, ex recensione I. Bekkeri, edidit Academia Regia Borussica, Vol. I; Berolini: Reimer.

Arntzenius, F. (2000). Are there really instantaneous velocities?, *The Monist* 83, 187-208.

Arntzenius, F. (2012). *Space, Time and Stuff*. Oxford: Oxford University Press.

Arntzenius, F., and Hawthorne, J. (2005). Gunk and continuous variation, *The Monist* 88, 441-466.

Arsenijević, M. (2003). Generalized concepts of syntactically and semantically trivial differences and instant-based and period-based time ontologies, *Journal of Applied Logic* 1, 1-12.

Arsenijević, M., and Kapetanović, M. (2008a). The "great struggle" between Cantorians and Neo-Aristotelians: Much ado about nothing, *Gratzer Philosophische Studien* 76, 9-90.

Arsenijević, M., and Kapetanović, M. (2008b). An $L_{\omega_1\omega_1}$ axiomatization of the linear Archimedean continua as merely relational structures, *WSEAS Transactions on Mathematics*, Issue 2, Volume 7, 39-47.

Arsenijević, M., Šćepanović, S., and Massey, G. (2008). A new reconstruction of Zeno's "Flying Arrow", *Apeiron* 41, 1-43.

Benthem, J. van (1991). *The Logic of Time*. Dordrecht: Kluwer.

Benthem, J. van (1995). Temporal logic, in: *Handbook of Logic in Artificial Intelligence and Logic Programming*, Gabbay, D. M., Hogger, C. J. and Robinson, J. A. (editors), Vol. 4. Oxford: Clarendon Press, 241-350.

Bochman, A. (1990a). Concerted instant-interval temporal semantics I: Temporal ontologies, *Notre Dame Journal of Formal Logic* 31, 403-414.

Bochman, A. (1990b). Concerted instant-interval temporal semantics II: Temporal valuations and logics of change, *Notre Dame Journal of Formal Logic* 31, 580-601.

Boyer, C. B., and Marzbach, U. (2011). *A History of Mathematics* (third edition). New York: Wiley.

Burgess, J. P. (1982). Axioms for tense logic II: Time periods, *Notre Dame Journal of Formal Logic* 23, 375-383.

Cantor, G. (1962). *Gesammelte Abhandlungen: Mathematischen und philosophischen Inhalts.* Hildesheim: Georg Olms.

Carnap, R. (1928). *Der logische Aufbau der Welt.* Leipzig: Felix Meiner.

Clarke, B. (1981). A calculus of individuals based on "connection", *Notre Dame Journal of Formal Logic* 22, 204-218.

Comer, S. C. (1985). The elementary theory of interval real numbers, *Zeitschrift für Mathematische Logik and Grundlagen der Mathematik* 31, 89-95.

Diels, H., and Kranz, W. (cited as *DK*) (1974). *Die Fragmente der Vorsokratiker.* Dublin/Zürich: Weidmann.

Earman, J., and Roberts, J. (2006). Contact with the nomic: a challenge for deniers of Humean supervenience about the laws of nature, *Philosophy and Phenomenological Research* 71, 253-286.

Foldes, S. (1980). On intervals in relational structures, *Zeitschrift für Mathematische Logik and Grundlagen der Mathematik* 26, 97-101.

Fränkel, H (1942). Zeno of Elea's attacks on plurality, *American Journal of Philology 63*, 193-206.

Grünbaum, A. (1952). A consistent conception of the extended linear continuum as aggregate of unextended elements, *Philosophy of Science* 4, 288-306.

Grünbaum, A. (1974). *Philosophical Problems of Space and Time.* Dordrecht: Reidel.

Hamblin, C. L. (1969). Starting and stopping, *The Monist* 53, 410-425.

Hamblin, C. L. (1971). Instants and Intervals, *Studium Generale* 24, 127-134.

Hilbert, D. (1902). Mathematical problems, *Bullettin of the American Mathematical Society* 8, 437-479.

Hilbert, D., and Ackermann, W. (1968). *Principles of Mathematical Logic.* Providence: Chelsea.

Hodges, W. (1993). *Model Theory.* Cambridge: Cambridge University Press.

Humberstone, I. L. (1979). Interval semantics for tense logic: some remarks, *Journal of Philosophical Logic* 8, 171-196.

Lewis, D. (1991). *Parts of Classes*. New York: Wiley-Blackwell.

Lewis, D. (1994). Humean supervenience debugged, *Mind* 103, 473-490.

Meredith, C. (1954). Single axioms for the systems (C,N), $(C,0)$ and (A,N) of the two-valued propositional calculus, *Journal of Computing Systems*, 155-164.

Mormann, T. (2000). Topological representations of mereological systems, in: *Things, Facts and Events*, Faye, J. (editor). Rodopi, 463-486.

Munkres, J. (2000). *Topology* (second edition). New Jersey: Prentice-Hall.

Needham, P. (1981). Temporal intervals and temporal order, *Logique et Analyse* 24, 49-64.

Quine, W. O. (1961). *From a Logical Point of View*. Harvard: Harvard University Press.

Robinson, D. (1989). Matter, motion and Humean supervenience, *Australasian Journal of Philosophy* 67, 394-409.

Roeper, P. (1997). Region-based topology, *Journal of Philosophical Logic* 26, 251-309.

Roeper, P. (2006). The Aristotelian continuum: A formal characterisation, *Notre Dame Journal of Formal Logic* 47, 211-232.

Russell, B. (1903). *The Principles of Mathematics*. London: George Allen & Unwin.

Russell, B. (1914). *Our Knowledge of the External World*. London: George Allen & Unwin.

Salmon, W. C. (1975). *Space, Time and Motion*. New Jersey: Dickenson.

Sambin, G. (2003). Some points in formal topology, *Theoretical Computer Science* 305, 347-408.

White, M. J. (1988). An "almost classical" period-based tense logic, *Notre Dame Journal of Formal Logic* 29, 438-453.

THIS MOMENT AND THE NEXT MOMENT

FRANCESCO ORILIA
Università di Macerata
Dipartimento di Studi Umanistici—Lingue, Mediazione, Storia, Lettere, Filosofia
orilia@unimc.it

ABSTRACT. This paper outlines a version of instantaneous presentism, according to which the present is a point-like instant, and defends it from two prominent objections. The first one has to do with the difficulty of accounting, from the point of view of instantaneous presentism, for the existence of events that take time, *dynamic* events, which cannot be confined to a single instant. The second objection is of a Zenonian nature and arises once time is viewed as a continuum that can be subdivided *ad infinitum*. In response to the first problem, the instantaneous presentist can appeal to past and future instants as long as it is assumed that there is no event that occurs at them. Once they are granted, one can then appeal to "past-oriented" and "future-oriented" properties such as *was P at t* (or *will be P at t*), where t is a past or future instant, in order to view dynamic events as occurring at the present instant. In response to the second problem, the instantaneous presentist could either try to digest the difficulties inherent in the view that time is discrete or accept the idea that, as a new instant becomes present, there is no preceding instant which is immediately past; there would only be infinitely many past instants with increasingly lower degrees of pastness.

1. Introduction

Very roughly speaking, presentism is an A-theory of time (a theory that acknowledges objective A-determinations or properties such as present, past and future), according to which only what is present is objectively real. It is typically assumed that presentism takes the present to be durationless. It often seems as if presentism with a durationless present, *thin presentism*, as we may call it (following Hestevold 2008), is the only presentist option on the market (see, e.g., Dainton 2010). But in fact there are philosophers who have defended, or at least considered, what we might call (following McKinnon 2003) *durational presentism*, or (following Hestevold 2008) *thick presentism*, according to which the present has some duration. McKinnon 2003 considers it, but does not endorse it. Hestevold 2008 considers two versions of it, *unlimited thick presentism*, according to which the present can be indefinitely long, a day, a year or even a millenium,[1] and *limited thick presentism*, according to which the present has a very short duration, so short that it has room only for extraordinarily brief events such as a butterfly's flapping its wings twice (Hestevold 2008, 334). Hestevold also distinguishes between those presentists who admit times as primitive entities, *timed presentists*, and those that do not, *time-free presentists*. Thin time-free presentism is arguably the default version of presentism. Thin timed Presentism, *instantaneous presentism* in Craig's (2000) terminology, is instead less popular. Nevertheless, in my view, it has some attractive features and is worth being taken seriously. At least according to its recent critics, Hestevold and Craig, there are, however, two serious obstacles in its way (setting aside other less urgent problems). One has to do with the fact that the present, *this* moment, *seems* to us to have some duration to the extent that we *observe* change, e.g., a moving object; it is because of this that Hestevold 2008 favors thick over thin presentism. The other is a Zenonian difficulty, on which Craig 2000 has insisted: instantaneous presentism takes the present moment to be a point-like instant, but then how can this moment ever be followed by subsequent moments, given that time had better be viewed as continuous and thus in such a way that there are infinitely many point-like elements in between two given moments?

In the following I would like to address these obstacles. Before doing this, I shall characterize instantaneous presentism, as I see it, in some more detail and hint at some of the reasons that militate in its favor. Some other reasons will emerge in discussing the first obstacle.

2. *Instantaneous presentism*

Typically, in being time-free presentists, presentists have pursued a Prior-style reduction of times to something like maximal and consistent propositions (Bourne 2006, Crisp 2007). Given this strategy, times exist only as abstract entities, just like numbers. As times are propositions, to assert that something occurs at a time amounts to saying that a certain proposition is entailed by another proposition. More importantly for present purposes, to say that a time is past or future is simply to say that a certain proposition

---

[1] Hestevold (2008, 332) hints that Craig and Merricks hold this view, although he refrains from taking a definite stand on this attribution. McKinnon 2003, on the other hand, attributes to Craig the view that the very issue of the extent of the present is ill-founded. For our purposes we need not inquire on this.

was true and that another proposition *will be* true. There is, in other words, according to this strategy, no real commitment to past and future times, but only to propositions that were true and propositions that will be true.

It is not clear, however, that these reductions can be consistently and non-circularly carried out (Craig 2000, 213, Oaklander 2010). If no appropriate reduction is forthcoming, it seems we are left either with a reduction of times to events or with times as primitive entities. The first option is, I would say, hardly compatible with presentism, for it will presumably imply a commitment to past and future events. The other leads to timed presentism. The primitive times (or moments) of this second option should be viewed as arranged in a sequence or succession, the "time series" (or more simply "time"), ordered by the asymmetric B-relation *earlier*, which we'll symbolize with "<" (the converse of this relation, *later*, can then be symbolized with ">"). Moreover, times are such that events can *occur* (*take place*) *at* them. The time series can be conceived of, metaphorically, as an empty container into which events can be placed, but that exists independently of what is placed in it, as in the "absolutist" or "substantivalist" view of time usually attributed to Newton (Markosian 2010, §2).

At first glance, timed presentism appears to be committed to only one instant, the present one, and to the "appearance and disappearance of one *time* after another" (Hestevold 2008, 328), for otherwise it should acknowledge past and future times and thus entities, the times in question, that are past or future and thus not present. As I see it, however, the timed presentist can allow for times that coexist with the present time and thus are past or future, since the sense in which the times other than the present are past or future is not one that should bother the presentist. For they are past or future not in the sense that they existed or will exist, but fail to exist (now). They exist, one can say, *sic et simpliciter*, just like the present time and just like abstract entities such as numbers. Or, perhaps, alternatively, they exist at the present time just like the present time itself. What differentiates them from the present time is that only the latter is such that events occur at it. The former instead are "empty," no event occurs at them, for each event occurs at the present time.[2]

---

[2] A referee has put forward the worry that such empty times can "hardly be of any service for any real problem," since they should turn out to be all identical to one another and such that it is impossible to refer to any of them, given that they all lack "any physical content." Here it should be kept in mind that according to the substantivalist perspective on time, times differ from one another in that they essentially occupy different positions in the time series and their existence should be regarded as indepen-

Intuitively, events can be of two kinds: *static*, e.g., a cat's being on a mat, or *dynamic*, the running of a rabbit (Casati and Varzi 2010). In a derivative sense, we can say that the objects involved in an event that occurs at a certain time, e.g., the cat and the mat of our first example, also occur at that time. We might add that, in addition to these *ordinary* events, there also occur at the present time t, *chronological* events. That is, for any time t' < t, events such as t's being past, and, similarly, for any time t' > t, events such as t's being future.

As noted, instantaneous presentism is a variety of timed presentism. According to it, the moments or times of the time series at which events take place are point-like or durationless: they are moments that involve neither an earlier "part" (a sub-moment classifiable as past in an A-theoretic perspective) nor a later "part" (a sub-moment classifiable as future in an A-theoretic perspective). In a word, they are *instants*. Intervals stretching from one instant to another of the time series may also be called *times* or *moments*, but the term "instant" will not be used for them).

---

dent from our ability to refer to any of them in particular. It can be pointed out, however, that tensed properties may further differentiate them and offer ways to refer to them. For instance, it is true of a certain time, t_1, but not of another time, t_2, that there *was* a certain event, e.g., the beginning of the battle of Waterloo, that occur*red* at t_1, but not at t_2. Thus, even though the time at which the battle of Waterloo started is now empty, it can be still referred to by the definite description "the time at which the battle of Waterloo started." On behalf of the theoretical services for which times as so conceived can offer, note that, in spite of the fact that the opposite "relationist" view of time (often attributed to Leibniz) is apparently more popular, some still find the substantivalist picture more appealing, because, e.g., it makes room for the intuition that in principle there may be periods of time in which nothing happens. This intuition has been strengthened by Shoemaker (1969), who has shown that under certain circumstances one could even come to know the length of these periods (Markosian 2010, §2). Moreover, the times of the substantivalist conception are, it seems to me, implicitly appealed to in certain proposals advanced by presentists in order to deal with the truthmaking objection (Bigelow 1996, Keller 2004). For example, Keller (2004, §3.1) proposes that, in order to provide a truthmaker for the true sentence "Anne Boleyn was executed at around midday on May 19, 1536," one may point to the fact that the world as a whole still has the property of being such that at a certain specific time (midday, May 19, 1536, or something in the vicinity) Anne Boleyn was executed. Here, however, we shall focus on other issues with respect to which the times of the substantivalist perspective can serve a theoretical purpose.

3. This moment

The first obstacle for instantaneosus presentism that I want to consider is based on the apparent existence of dynamic events. They seem to "occupy" an extended time and thus seem to occur at times characterizable as intervals, involving an earlier and a later part, rather than at instants. This seems testified by the fact that we *perceive* dynamic events. To put it otherwise, our experienced or specious present seems to involve an earlier and a later part.

Consider for example the dynamic event of a body, say a billiard ball, moving from a place to another, an event characterizable as follows (I use asterisks to refer to events and sometimes also to refer to properties or relations):

(1) *x moves from p_1 to p_2*.

This event, it seems, takes place, not just at an instant, but at a certain interval, the interval T, which goes, say, from time t_1 to time t_2. If so, it seems plausible to say that this event somehow involves the static events

(1a) *x is in p_1*,

occurring at t_1

and

(1b) *x is in p_2*,

occurring at t_2.

A subject that observes this event has an impression of movement, say as of a moving ball, in its specious present, which we should also take, one might urge, as occurring at the interval T, rather than at a single instant.

Now, to the extent that we take this at face value, we clearly have a problem for instantaneous presentism, for according to it only a single instant can be present. The thick presentist, in contrast, can regard the whole T as present. It should be noted however that this comes with a price: thick presentism must sacrifice an intuitively valid principle, the *precedence principle*. This tells us that an event e that precedes, or is before, another (present) event e' is past. To see why, imagine that we are observing the event (1) and thus we are inclined to consider it as existent. From the presentist's perspective, to the extent that this event exists, we must regard T as present. This in turn should incline us to consider both (1a) and (1b) as present. On the other hand, (1a) must be taken to occur before (1b). And thus the presentist must sacrifice the precedence principle. For otherwise she will have to hold that (1a) is both present and past, in conflict with what we could call *McTaggart's Principle*, according to which A-properties such as present and past are incompatible, at least inasmuch as they are applied to events.

Can instantaneosus presentism be reconciled with the existence of dynamic events and thus provide a version of presentism that does not sacrifice the Precedence principle or McTaggart's principle? I think it can. Let us see how by focusing on event (1). What the instantaneous presentist can say is that he movement of the billiard ball x consists in the fact that at instant t_1 it had the property *is in p_1* and another property, to be considered in a moment, that has to do with its position at the successive instant t_2. Moreover, at time t_2 it has, *in addition* to the property *is in p_2*, the "time-indexed" property *was in p_1 at time t_1*. The idea is that, although only the present time is replete with objects and static events, in such present time there may occur objects with properties indexed with past times (*past-oriented* properties), i.e. static events consisting of objects' having such time-indexed properties. Furthermore, there may also at play at the present time other time-indexed properties having to do with the future of the object in question (*future-oriented* properties, in that they are indexed with a future time). To clarify this, let us go back to our event (1). We said that at time t_1 x had the property *is in p_1* and another property having to do with its position at the successive instant t_2. What is this property? We can understand it in two ways, depending on whether or not we take it as fully determined at t_1 that x would have then been in p_2 at time t_2. If it was fully determined, then x had at time t_1 the time-indexed property *will be in p_2 at time t_2* (and, correspondingly, it has at t_2 the property *was at t_1 such that it would have been at p_2 at t_2*). On the other hand, if we do not take it to be fully determined at t_1 that x would have been in p_2 at time t_2, the indexed property concerning its future that x had at t_1 was a mere *propensity*, a property such as *x is potentially at place p_2 at time t_2* (and, correspondingly, x has at t_2 the property *was at t_1 potentially in p_2 at time t_2*). We can conceive of the property *x is potentially at place p_2 at time t_2* as a property that x has insofar as, roughly speaking, it had stored (say, by having been pushed) some kinetic energy leading in a certain direction. Having this property does not necessarily result in being in p_2 at time t_2, for, e.g., there can be an intervening obstacle.[3] In sum, according to this perspective, a dynamic event is, we may say, something that supervenes on static events involving these time-indexed properties, which occur at a certain instant, such as *x was in p_1 at t_1*, *x at t_1 was potentially in p_2 at t_2*, *x is in p_2*. Or, if one wishes, a dynamic event is a "conjunction" of such static events (a conjunctive event). In any case, the dynamic event can be said to occur at an interval, say the interval T going from t_1 to t_2, only in a derivative sense based on the fact that the event involves time-indexed properties such that t_1 and t_2 are the indexes.

Thus, for example, at the interval T, there occurs the dynamic event (1), insofar as at t_2 there occurs this conjunction of static events

(1')   *x was in p_1 at t_1, was at t_1 potentially in p_2 at t_2, and is in p_2*.

What can we say from this perspective about the perception of movement, the presence, so to speak, of a moving object in the specious present? Clearly, from the point of view of instantaneous presentism, the specious present of a given subject must occupy a single instant and thus is not temporally extended, despite the fact that it provides the subject with an impression of becoming, of something (e.g., event (1b)) coming after something else (e.g., event (1a)). Although the specious present of instantaneous presentism must occur at an instant, it may well involve the perception of a dynamic event to the extent that events are conceived of in the way we have seen. Consider (1) again. At t_2 an observer in the vicinity of p_1 may see the dynamic event (1) in the sense that she observes static events such as these (assuming the non-determinist option): *x was in p_1 at t_1*, *x was potentially in p_2 at t_1*, *x is in p_2* and (let us suppose) *x is potentially in

---

[3] Propensities might seem to someone rather obscure entities belonging to an outdated Aristotelian metaphysics. But in fact properties of this kind are postulated and characterized in some detail by Lange (2005) in order to make sense of instantaneous velocity in the context of our current scientific outlook (see Dainton 2010a, §17.2 for a discussion). Harrington (2008) analyzes Lange's approach and then defends another view. But Harrington also acknowledges that he has no knock-down arguments against Lange's view. In the light of works such as these, propensities are not so obscure after all and, independently of the problem under consideration here, there may well be good reasons to appeal to them, namely to account for instantaneous velocity.

p_3 at t_3*[4] (we neglect here for simplicity's sake the fact that in a causal account of perception there is a time lag between an event and the perception of the event). We shall discuss in a moment in what sense one can be said to "observe" events with past-oriented or future-oriented properties.

As Dainton 2010 notes, thin presentism (presentism *tout court*, according to Dainton) is compatible with two accounts of the specious present: the kinematic model and the retentional model. According to the former, it is simply the rapid succession of single episodes of consciousness with different contents that gives us the impression of becoming. According to the latter, a single episode of consciousness, even if occurring at a single instant, is by itself sufficient to give us an impression of becoming, for it involves, in Husserl's terminology, not only "primal impressions" of what is presently going on, but also "retentions" of the immediate past and "protentions" of the immediate future. As Husserl sees it, primal impressions do not altogether vanish, but somehow survive for a very short while as retentions, i.e. as distinctive forms of consciousness that accompany new primal impressions, so as to make it seem that there is change. To reach this effect, the primal impressions are also typically accompanied by protentions, characterizable as anticipations or expectations that something is going to happen, and that can be more or less detailed, depending on the extent to which what one is observing is of a familiar kind (see Dainton 2010, § 2.7). For reasons well expressed by Dainton, I think that the retentional model is much more promising, and, interestingly, the account of dynamic events given here points in its direction. For we can say, very roughly speaking, that (in typical cases of veridical perceptions) retentions and protentions in our specious present correspond, respectively, to the exemplification by objects in outer reality of past-oriented and future-oriented time-indexed properties and relations. Thus, one can say that one observes such exemplifications inasmuch as one has appropriate retentions and protentions. For instance, to observe (at t_2) that x was in p_1 at t_1 and that it was at t_1 potentially in p_2 at t_2 is to have a Husserlian retention (at t_2), and to observe (at t_2) that x is potentially at p_3 at t_3 is to have a Husserlian protention (at t_2). It should be noted that experiencing a protention is no guarantee that the expectation that it amounts to comes to be fulfilled. One experiences a protention, it is being suggested, in response to an object that exemplifies a propensity, e.g., x, which exemplifies at t_2 a property such as being potentially in p_3 at t_3. But the exemplification of a propensity does

---

[4] Or, if one wishes, what is observed is a conjunctive event involving as conjuncts *x was in p_1 at time t_1*, *x was at t_1 potentially in p_2 at time t_2*, etc.

not in itself guarantees that the propensity will be realized: x may fail to reach p_3 (as noted, there may be, e.g., an unexpected intervening obstacle, say, a transparent screen).[5]

4. The next moment

Let us now move to the second obstacle. Traditionally, these two options regarding time are in play:

DT. *Discreteness of Time*. Time is discrete, i.e., for any moment t_1 there is an immediate successor, the *next* element t_2, i.e., an element t_2 such that $t_1<t_2$ and there is no t such that $t_1<t<t_2$. In other words, the time series is a succession of *consecutive* items.

CT. *Continuity of Time*. Time is continuous and thus in particular it is dense, i.e., if t_1 and t_2 are two moments such that $t_1<t_2$, then there is another moment t such that $t_1<t<t_2$. In other words, the time series is a succession of items without consecutiveness: for no element in the series there is an immediate successor.

From the point of view of instantaneous presentism, for there to be temporal passage or becoming, presentness must shift, so to speak, from one instant to another. In other words, the present instant must be able to loose presentness and correspondingly another instant must acquire it. But depending on whether we accept DT or CT, the matter looks quite different. Let us look at both options in turn.

4.1. Is time discrete?

It is intuitively simple to conceive of how the present instant looses presentness and another acquires it, if we accept DT, so that, for any instant, there is the next instant. If this is the case, presentness will just shift from this instant to its immediate successor, the next instant. Moreover, we can take at face value the fact that we normally use the expression "the next moment." For example, we say things such as "when John realized he had run out of water, he felt hopeless, but the next moment found him relieved

---
[5] For more details on the approach proposed in this section see Orilia 2012.

as he spotted a fountain." Last but not least, given DT, we can more easily accommodate, it would seem, what some physicists tell us, namely that quantum mechanics should incline us to accept that there are minimal discrete quantities of time, the so-called "chronons" (Whitrow 1980, 203, Smolin 2000, 2006).

If we take this line, there is the obvious difficulty that in describing the world and in doing physics, at least at the macroscopic level, we seem to need real numbers in a way that suggests that time is continuous (Dainton 2010a, 271). For example, if a physical square has a 1 meter long side, we must say that its diagonal is √2 meter long. Consider then a body moving at the uniform speed of 1 meter per second. It will take, it seems we should say, √2 seconds to move along the diagonal. The example suggests that we need reals to measure time, which in turn suggests that time is continuous (Salmon 1970a, 35).

This latter difficulty is perhaps not insurmountable, for the experts tell us that there are systems of discrete mathematics that could be used to do physics, and that we could in principle view the use of reals as an approximation of what we should really do (Caratheodory 1963, Penrose 2004, Ch. 16, Fano et al. 2009). But there is another serious hurdle, namely Zeno's paradox of the stadium, at least as it is understood by Russell 1914. According to Owen 1957-58, Zeno wanted to show that both the hypotheses that time and space are continuous and that they are discrete lead to trouble. The paradox of the Stadium and the plurality paradoxes were designed for the former task and the more well-known puzzles, such as the arrow and race paradoxes, were designed for the latter task. The historical accuracy of this interpretation of Zeno may be questioned (see, e.g., Sorabji 1983, 331), but, independently of this, I am inclined to say that the paradox of the stadium can indeed be understood in a way that creates a problem for DT (in line not only with Russell 1914, but also with Grünbaum 1968 and, more recently, Craig 2000, Ch. 7, and Dainton 2010a, § 17.3). Let us see how.

We know the paradox of the stadium from Aristotle's *Physics* (VI, 9, 293b). As Aristotle presents it, it may seem nothing more than a fallacious piece of reasoning that naively neglects the notion of relative speed (Salmon 1970a, 11). Nevertheless, Russell 1914, relying on Gaye 1910, presents it as directed against DT as follows.[6] Imagine this situation. There

---

[6] Owen (1957-58, 149) interprets the puzzle more or less like Russell except that he concentrates on space and sees it as directed against the assumption that there are mi-

is a series of consecutive places, individuated by their being occupied by three adjacent bodies, A_1, A_2, A_3, of equal size. Call "point" the size in question. Thus, we can say that each A occupies one point. Imagine further that two series of bodies of the same size, B_1, B_2, B_3, and C_1, C_2, C_3,, each occupying one point, are aligned at a certain instant t_n, respectively above and below the A's. We thus have the following situation N at the instant t_n:

Situation N, at instant t_n

$B_1\ B_2\ B_3$
$A_1\ A_2\ A_3$
$C_1\ C_2\ C_3$

Imagine that the B-bodies and the C-bodies are travelling in opposite directions at the absolutely minimum speed of one point per instant, while the A-bodies stand still. Hence, after one instant, B_2 has gone from being above A_2 to being above A_1 and C_2 from being below A_2 to being below A_3. We then have the following situation $N+1$, at instant t_{n+1}:

Situation N+1, at instant t_{n+1}

$\Leftarrow\quad B_1\ B_2\ B_3$
$\quad\quad A_1\ A_2\ A_3$
$\quad\quad\quad C_1\ C_2\ C_3\quad \Rightarrow$

Now, given DT, we have a problem. C_1 is (from the reader's perspective) on the left with respect to B_2 at instant t_n and on the right at instant t_{n+1}, and thus intuitively there should be an intermediate situation and an intermediate time $t_{1'}$ at which C_1 and B_2 are aligned (each being in a smaller point, respectively above and below a smaller point in between A_1 and A_2):

Hypothetical situation N+1', at hypothetical intermediate time $t_{1'}$

$\Leftarrow\quad B_1\ B_2\ B_3$
$\quad\quad A_1\ A_2\ A_3$
$\quad\quad\quad C_1\ C_2\ C_3\quad \Rightarrow$

---

nimal units of space ("infinitesimal quantities"). For a simplified version of the problem, see Salmon 1975, 64-65.

However, given DT, t_n and t_{n+1} are consecutive and thus there cannot be this intermediate time. As Russell (1914, 182) puts it: "When then did *B* [B_2] pass *C'* [C_1]? It must have been somewhere between the two moments which we supposed consecutive, and therefore the two moments cannot really have been consecutive. It follows that there must be other moments between any two given moments and therefore there must be an infinite number of moments in any given interval of time."

This argument might be taken to beg the question against the supporter of DT, who after all could claim that there is never a moment at which B_2 passes C_1; counterintuitive as this might seem, there is just a leap, so to speak, from *N* to *N*+1, and thus no intermediate situation (Salmon 1975, 65, Whitrow 1980, 191, Dainton 2010a, 296).[7] But even if the argument is not conclusive, it points out that DT sacrifices a very strong intuition, for after all the inclination to believe that at some time B_2 and C_1 must be aligned is hard to resist. This is a price that one may not want to pay.[8]

4.2. Is time continuous?

In the light of the difficulties discussed in the previous subsection, one may want to accept CT rather than DT. But is this option open to the instantaneous presentist? According to Craig (2000), it is not. As we shall see, Craig's attack is not novel, because a similar point has been made by renowned thinkers such as James, Whitehead (as acknowledged by Craig himself) and Findlay.

4.2.1. Craig's attack

Here is how Craig (2000, 231) formulates the problem:

> How can time progress instant by instant, since instants have no immediate successors? Like Achilles in the Dichotomy Paradox, the present instant would be frozen into immobility; it could not elapse to be succeeded by another, for before any new instant

---

[7] It must be granted of course that the A-, B- and C-bodies in question operate at such a fine-grained level that no observation could in principle tell us whether or not there is such a leap.

[8] Another difficulty for the supporter of DT is Weyl's so called tiling problem, but even this one can be circumvented (Dainton 2010a, 297).

could be actualized an infinite number of prior instants would have to be actualized first. And even if the present instant could elapse, how could any non-zero interval of time elapse, since the addition of durationless instants will never add up to the lapse of a single nanosecond of time? But if no non-zero interval of time ever elapses, temporal passage is impossible.

It is important to note that, before this passage, Craig contends (2000, 230) that the present cannot be regarded as the boundary of two intervals, one containing the past instants, i.e. all instants before the present one, and another containing all future instants, i.e., all instants after the present instant. For, as Craig understands instantaneous presentism, only the present instant is actual and thus, *a fortiori*, such intervals cannot also be actual, for they are made up of past and future instants and these are not actual. Thus, says Craig, according to instantaneous presentism, there can be at most one "degenerate interval," i.e., an interval of zero duration, namely, the present instant. It follows that temporal becoming must be conceived of as "the progressive actualization of new instants."

To see what Craig has in mind, it is also important to focus on these two passages (in Craig 2000, pp. 235 and 236, respectively):

> [...] let us suppose [...] that the basic parts of time are instants. In order for the present instant to elapse it must be succeeded by another. But there is no immediate successor to the present instant. Before the succeeding instant can become present, an infinite number of succeeding instants will have to become present first. The point is not the fallacious inference that it would therefore take an infinite amount of time for the succeeding instant to become actual. Rather the point is that no instant can succeed the present instant. For no instant can immediately succeed the present. The present would therefore exist as the *nunc stans* of the classical doctrine of eternity, not the *nunc movens* of A-theoretic time.

> Not only does the identification of the present with a degenerate interval of time thus fall prey to Zeno's paradoxes of motion, but moreover, it also falls into the snare of Zeno's paradoxes of plurality. [...] If we conceive temporal becoming to proceed instant by instant, the length of time between some past event or moment and the present could never increase, since the lapse of durationless instants adds nothing to the interval between the past instant and the present instant. But then there is no "flow" of time at all, and we are left again with the *nunc stans* of the present instant,

never able to recede into the past. The doctrine of the instantaneous present is thus incompatible with objective temporal becoming.

As is evident from these quotations, there are two problems.[9] One has to do (loosely speaking, as we shall see) with Zeno's paradoxes of motion and the other with Zeno's paradox of plurality. Let us first focus on the former.

Actually, Craig is not really exploiting Zeno's paradoxes of motion, which have to do with both time and space. Rather, following the lead of James (1911, 181) and Whitehead (1925, 127; 1929, 94), he is trying to take advantage of an argument that, although derived from Zeno's paradoxes about space and time (those designed to show that the hypothesis of the continuity of time and space leads to trouble), makes no reference to space: a "distilled Zenonian argument (based on CT)," as we may call it, in which there is no essential reference to space. Here is Whitehead's version (1929, 94):

> The [distilled Zenonian] argument [based on CT], so far as it is valid, elicits a contradiction from the two premises: that in becoming something (*res vera*) becomes and (ii) that every act of becoming is divisible into earlier and later sections which are themselves acts of becoming. Consider, for example, an act of becoming during one second. The act is divisible into two acts, one during the earlier half of the second, the other during the later half of the second. Thus that which becomes during the whole second

---

[9] The first passage may be misleading, because it could give the impression that Craig falls prey to a simple fallacy. When he affirms that "no instant can immediately succeed the present" as a reason for the claim that "no instant can succeed the present instant," one might be tempted to reply, as pointed out by a referee, that by the same token "no real number can be greater than another real number because no real number is *immediately* greater than another (i.e., greater without there being other numbers between them)" (here I quote the referee). But it is not obvious that this analogy holds, because we think of the real numbers as already given in one fell swoop, as members of an actually infinite set ordered by the 'greater than' relation; no becoming is involved in their case. In contrast, Craig is not thinking of all the instants as already given. No more than one instant can be actual, the present instant, and this must fade into non-actuality as another one *becomes* present. The problem has to do specifically with *becoming*, as we shall see in more detail by also considering similar worries by Whitehead and Findlay (the quotation from Whitehead provided below is especially explicit in its focusing on becoming). Thus, we cannot simply put the matter to rest with an analogical counterargument that points to a sequence, that of real numbers, that involves no becoming.

presupposes that which becomes during the first half-second. Analogously, that which becomes during the first half-second presupposes that which becomes during the first quarter-second, and so on indefinitely. Thus if we consider the process of becoming up to the beginning of the second in question, and ask what then becomes, no answer can be given. For, whatever creature we indicate presupposes an earlier creature which became after the beginning of the second and antecedently to the indicated creature. Therefore there is nothing which becomes, so as to effect a transition into the second in question.

Findlay (1941, 156-157) puts it this way:

> when we strip Zeno's problem of its spatial and other wrappings, its significance becomes clearer. For it is not, essentially, a problem of space or quantity, but solely of time [...] the problem assumes its most vexing form if we allow that ordinary happenings have ultimate parts that take no time. For of such parts *it seems most natural to say that none can be next to any other* [italics mine], and once this is said it is hard to understand how any ultimate part can ever pass away or be replaced by any other. For before such part can be replaced by any other similar part, it must have first been replaced by an infinity of other similar parts. Our admission seems to leave us with a world immobilized and paralyzed, in which every object and process, like the arrow of Zeno, stands still in the instant, for the simple reason that it has no way of passing on to other instants.

In the above quotations Craig provides his own version of a distilled Zenonian argument, which is meant to be fatal for instantaneous presentism. It focuses on instants rather than on the "creatures" or "acts of becoming" of Whitehead's version and the "happenings" of Findlay's version. It seems to me that, without leaving out anything essential, Craig's version can be summarized as follows: For there to be becoming, the present instant, t_0, must elapse, i.e., as I understand Craig, cease to be actual, and a new instant must become present, i.e., actual. For if t_0 elapses without a new instant that becomes actual there would be no present, and if a new instant becomes actual and t_0 does not elapse there would be two present instants. In other words, t_0 cannot cease to be actual, unless one specific new instant becomes actual. But there is no candidate for this role of "new actual instant" that can succeed in filling in the role, given CT. The candidates, according to Craig, are non-actual possibilia arranged in a continuum by the relation <. One of them must become actual for the present instant to cease to be actual. But, given CT, any candidate instant t for this role is superseded by another candidate instant t' such that $t_0<t'$ and $t'<t$. And thus none of them can succeed in filling in the role. Hence, the present instant cannot cease to be actual and there is no becoming.

As regards Craig's attack on instantaneous presentism based on the paradox of plurality, it can be exhaustively summarized thus: we should admit not only that durationless instants elapse, but also that intervals elapse. But, even if, per impossible (given the previous difficulty) durationless instants could elapse, no interval could elapse. We could admit that an interval elapsed, if elapsed instants could "add up" to an interval, but we cannot do this, because it would mean inappropriately to regard an interval, which has a duration, as a sum of durationless instants. Moreover, the interval between a given past instant p and the present instant should be constantly increasing. But, according to instantaneous presentism, this must be by the addition of new durationless instants, as they cease to be present, to the interval stretching from p. However, this adding up cannot result in a longer interval, because all the added up instants have zero duration.[10]

In reply, the instantaneous presentist should urge that not only the present instant, but all instants are actual. In other words, the non-present instants are not mere possibilia that need be actualized in order to become present. They are always actual. The difference between the present and the non-present instants lies not in the lack of actuality of the former, but in their "emptiness," for events occur only at the present instant. Thus, no passage from non-actuality to actuality of a certain instant is needed for the present instant to elapse, and Craig's version of the distilled Zenonian argument is thus disposed of.

---

[10] It might be thought here that this is a bad argument for the simple reason that it ignores that the intervals in question can be regarded as non-denumerable sets (that are also unions of an at most denumerable number of disjoint sets), which, as such, may have different positive lengths, even if composed of elements of zero magnitude. As is well known, such an appeal to non-denumerable sets is exploited by Grünbaum (1968, Ch. 3) precisely to deal with Zeno's paradox of plurality. As we shall see below, however, Craig thinks that this move is not available to the instantaneous presentist, because it presupposes an already given set of elements, whereas in instantaneous presentism instants are given one by one, as they *become* present, and thus there is no infinite non-denumerable set of point-like instants that can ever be viewed as having a positive length.

Similarly, the first part of Craig's attack based on the Plurality Paradox is eschewed. For the instantaneous presentist cannot agree that there is temporal passage in the sense that intervals elapse, if elapsing is understood as loss of actuality. According to instantaneous presentism, rather, there is becoming, or temporal passage, in the sense that presentness (being "occupied" by events) shifts from one instant to another in the before-to-after "direction." This can give rise, and we shall worry about this below, to another version of the distilled Zenonian argument based on CT. But no Paradox of plurality should bother the instantaneous presentist. She can simply regard any interval from t_1 to t_2, where these are past instants, as a past interval and thus, if one wishes, as an "elapsed interval." As regards the second part of the attack, the instantaneous presentist, following Grünbaum's lead (1968), can simply regard the length of intervals, not as resulting from the (arithmetic) addition of zero lengths, but in terms of the set-theoretical union of a continuum of degenerate intervals (of zero length). Craig thinks that this option is not available to the instantaneous presentist and that the only option available to her is arithmetical addition, because "a past interval is formed by the successive addition of instants as they elapse." (Craig 2000, 237.) But, as we have seen, the instantaneous presentist need not regard intervals as constructed one by one in this way. They are always given, because, as we have seen, all instants are equally actual.

4.2.2. *The distilled Zenonian argument and the shifting of presentness*

But can a version of the distilled Zenonian argument based on CT threaten instantaneous presentism in view of the fact that presentness shifts along the time series? If we understand "elapsing" as standing for the property that a time acquires when it ceases to be present,[11] then Craig's argument can be reformulated as follows.

For there to be becoming, the present instant, t_0, must cease to be present, and a new instant must be present. For, if t_0 ceases to be present without a new instant that is present, there would be no present; and, if a new instant is present without t_0 ceasing to be present, there would be two present instants. In other words, t_0 cannot cease to be present, unless one

---

[11] According to the view defended here, that a time ceases to be present amounts to its ceasing to host events, i.e., its becoming "empty." But we can neglect this, as far we are concerned here.

specific new instant is present. But there is no candidate for this role of "new present instant" that can succeed in filling in the role, given CT. The candidates are arranged in a continuum by the relation <. One of them must become present for the present instant t_0 to lose presentness. But, given CT, any candidate instant t for this role is superseded by another candidate instant t' such that $t_0 < t'$ and $t' < t$. And thus none of them can succeed in filling in the role. Hence, the present instant cannot cease to be present and there is no becoming.

What can the instantaneous presentist reply? From the point of view of instantaneous presentism, since all events occur at the present instant, even the event consisting of the ceasing to be present of a certain instant t, *if there is such an event*, must occur at the present instant. But then, presumably, the present instant is the immediate successor of instant t. For suppose it is not. Thus, there is an instant $t_{1/2}$ after instant t and before the present instant. If so, $t_{1/2}$, rather than t, is the instant with the property of ceasing to be present. Let us then focus on the option, according to which no instant has (at the present instant) the property of ceasing to be present.

Can it be maintained? At first glance it might seem that it cannot, since, as noted above, it seems to imply that there is more than one present moment: if there is no time that ceases to be present as another time acquires presentness, aren't we stuck with two present times? But in fact this does not follow. For even if there is no time at the present instant with the *absolute* property of ceasing to be present, it can still be the case that, at the present instant, every instant before it has the property of ceasing to be present with a certain degree d, so that it is not present. We can see the matter in a better light, if we reflect on the fact that the property of ceasing to be present, as conceived of in the argument, really is the property of being immediately past, i.e., past in such a way that no other instant can be, so to speak, less past than it is. But, obviously, to admit this property amounts to deny CT. CT is however compatible with properties of the kind "past with degree d," where d is any real number. For a very low d, we can say that the pastness in question is of a kind that we may want to classify as "immediate past," but there is no d, no matter how small, such that the instant t_d with the property of being past with degree d is *the* instant that is just past. For we can always find a lower degree of pastness d' such that another instant $t_{d'}$ has the property of being past with degree d' and thus is, so to speak, "more immediately past" than t_d.

4.2.3. Temporal passage and ontological dependence

It could be suspected however that there is a problem lurking behind, a problem that has to do with ontological dependence. Although this notion is complicated and many varieties of it should perhaps be distinguished (see Lowe 1998, 2008, Correia 2005, Cameron 2008, Paseau 2010), for present purposes we should simply record that an entity x *ontologically depends* on an entity y (and, conversely, y is *ontologically prior* to x) only if x cannot exist unless y also exists.

To see why ontological dependence is of interest here consider that, intuitively, the presentness of the present instant t_0 cannot exist unless the pastness with degree d of a given instant $t<t_0$ also exists. For if t had not acquired the property of being past with a certain degree d, it should presumably be present and we would have two present instants. Thus, it would seem, for any $t<t_0$, the presentness of t_0 ontologically depends on the pastness with a certain degree d of t. And, similarly, for any instant t that is past with degree d, its pastness with degree d ontologically depends on the pastness with a higher degree d' of any instant $t'<t$.

To state these intuitions in a more general fashion, let us say that $A(t)$ is the chronological event consisting of time t having a certain A-property of the kind we have just discussed. Thus, e.g., $A(t)$ could be the event that t is present or the event that t is past with a certain degree. With this terminology at hand, the above intuitions can be put as follows:

DPPA. *Dependence of the chronological Posterior on the chronological Prior from an A-theoretic perspective.* $t<t'$ if and only if $A(t')$ ontologically depends on $A(t)$.

Now, there appears to be strong resistance against ungrounded chains of ontological dependence (Lowe 1998, 158, Cameron 2008). In these chains, x depends on y, y on z and so on *ad infinitum*, without an ultimate ground that founds the chain. It is protested that, for there to be dependence, such ground must exist. As far as I can see from the literature on ontological dependence, there aren't similar explicit animadversions against chains of ontological dependence without *direct* dependence, where an entity x *directly depends* (*ontologically*) on an entity y if and only if there is no z such that x ontologically depends on z and z ontologically depends on y. Yet, I suspect that there is little inclination to accept them. For in such chains, even if there is a ground, there is a sense in which the ground can "never" be reached from a given element of the chain. This is so, because in these chains, if x ontologically depends on y, then there cannot be a finite dependence chain linking x to y, i.e., a finite sequence of entities $z_1, ..., z_n$, such that x directly depends on z_1, z_1 directly depends on z_2, ..., z_n directly depends on y. Thus, in these *dense* chains, as we may call them, it is as if the ground could not really exercise its grounding function, given its infinite distance from any element of the chain. Suppose then that one rejects dense chains of ontological dependence and buys the following:

NDWDD. *No Dependence Without Direct Dependence*. If x ontologically depends on y, then there must be a finite dependence chain linking x to y.

Once NDWDD is granted, CT and DPPA immediately lead us to a problem, since they force us to say that, if $A(t_2)$ depends on $A(t_1)$ (given that $t_1 < t_2$), then no finite dependence chain linking $A(t_2)$ to $A(t_1)$ can exist. If this chain existed, its nodes would obviously be entities of the type $A(t)$, where t is either t_1, t_2 or a time intermediate between t_1 and t_2. And whatever length we choose for the chain, we can derive a contradiction. Suppose, e.g., that the chain is of length zero, i.e. $A(t_2)$ *directly* depends on $A(t_1)$ (say, t_2 has a very low degree of futurity, so that it is just about to become present and t_1 is present). This cannot be the case because, by CT, there is a time $t_{1'}$ such that $t_1 < t_{1'}$ and $t_{1'} < t_2$. And thus, by DPPA, $A(t_2)$ depends on $A(t_{1'})$ and $A(t_{1'})$ depends on $A(t_1)$, in contrast with the hypothesis that $A(t_2)$ *directly* depends on $A(t_1)$. A similar reasoning applies, if we hypothesize that the chain is of length 1, so that, for a time $t_{1'}$ intermediate between t_1 and t_2, $A(t_2)$ *directly* depends on $A(t_{1'})$ and $A(t_{1'})$ directly depends on $A(t_1)$. And so on.

The instantaneous presentist who does not want to drop CT in favor of DT must drop NDWDD. I think that no real contradiction or absurdity follows, if we drop NDWDD. Since, however, some philosophers may not abandon it that easily, it is important to see that, once we assume it, problems arise not just for instantaneous presentism, but for any other theory of time that wishes to accept CT (or some counterpart of it). Let us illustrate this by focusing on a B-theory of time.

In constructing the argument, we cannot appeal to A-properties, but we can admit that, at any time t, the world is in a certain state. Let $t(w)$ be the state in which the world is at time t. With this terminology at hand, we can state this B-theoretical version of DPPA:

DPPB. *Dependence of the chronological Posterior on the chronological Prior from a B-theoretic perspective.* $t<t'$ if and only if $t'(w)$ ontologically depends on $t(w)$.

It should be clear that, with DPPB in place, we can, *mutatis mutandis*, unfold the above argument again and show that CT is at odds with either NDWDD or DPPB, just as it is at odds with either NDWDD or DPPA. Now, it seems to me, DPPB is, from the B-theoretical perspective, just as plausible as DPPA is from the A-theoretical perspective. And thus, the B-theorist who wants to retain CT had better reject NDWDD just like the instantaneous presentist who wants to retain CT.

5. Conclusion

We considered two objections against instantaneous presentism. In order to tackle the first one, an account of dynamic events compatible with this doctrine was proposed. According to it, the present, this moment, may well be a durationless instant. As regards the second objection, it was argued that, even if we take time to be continuous, no insurmountable difficulty peculiar to instantaneous presentism arises. If time is continuous, however, the instantaneous presentist cannot strictly speaking assert that this moment is followed by the next moment. In ordinary discourse, as noted, we do use the expression "the next moment." But the instantaneous presentist must claim that the expression in question should not be taken at face value. At best, it can be seen as referring to a certain subsequent time somehow selected by the context (Landman, 1991, 138-140, Bonomi and Zucchi 2001, 56-57).[12]

REFERENCES

Bigelow, J. (1996). Presentism and Properties. *Philosophical Perspectives*, 10, 35-52.

Bonomi, A., and Zucchi, A. (2001). *Tempo e linguaggio*. Milano: Bruno Mondadori.

---

[12] Thanks to Andrea Iacona, Vincenzo Fano and an anonymous referee for their comments on a previous version of this paper.

Bourne, C. (2006). *A Future for Presentism*. Oxford: Oxford University Press.

Cameron, R. (2008). Turtles All The Way Down: Regress, Priority And Fundamentality. *Philosophical Quarterly*, 58, 1-14.

Caratheodory, C. (1963). *Algebraic Theory of Measure and Integration*. New York: Chelsea Publishing Company.

Casati, R., and Varzi, A. C. (2010). Events. *The Stanford Encyclopedia of Philosophy (Spring 2010 Edition)*, ed. E. N. Zalta, <http://plato.stanford.edu/archives/spr2010/entries/events/>.

Correia, F. (2005). *Existential Dependence and Cognate Notions*. Munich: Philosophia Verlag.

Craig, W. L. (2000). *The Tensed Theory of Time*. Dordrecht: Kluwer.

Dainton, B. (2010). Temporal consciousness. *The Stanford Encyclopedia of Philosophy (Spring 2010 Edition)*, ed. E. N. Zalta, <http://plato.stanford.edu/entries/consciousness-temporal/>.

Dainton, B. (2010a). *Time and Space*, 2nd ed. Durham: Acumen.

Fano, V., Graziani, P., and Ridolfi, L. (2009). Movimento e velocità instantanea. *Thauma*, 3, 165-179.

Findlay, J. N. (1941). Time: A Treatment of Some Puzzles. *Australasian Journal of Philosophy*, 19, 216-235. Repr. in Gale 1968 (page references are to this version).

Gale, R. M. (ed.), (1968). *The Philosophy of Time*. New Jersey: Humanities Press.

Gaye, R. K. (1910). On Aristotle's Physics Z ix $239^{h}33$-$240^{n}18$. *Journal of Philology*, 31, 95-116.

Grünbaum, A. (1968). *Modern Science and Zeno's Paradoxes*. London: George Allen and Unwin LTD.

Harrington, J. (2009). Instants and Instantaneous Velocity. Presented at the 2009 Annual Meeting of The American Philosophical Association, New York, NY (abstract in the *Proceedings of The American Philosophical Association* of September 2009).

Hestevold, S. (2008). Presentism: Through Thick and Thin. *Pacific Philosophical Quarterly*, 89, 325-347.

James, W. (1911). *Some problems of philosophy*. New York: Longmans, Green and Co.

Keller, S. (2004). Presentism and Truthmaking. *Oxford Studies in Metaphysics, I*, ed. D. W. Zimmerman, Oxford: Clarendon Press, 83-104.

Landman, F. (1991). *Structures for Semantics*. Dordrecht: Kluwer.

Lange, M. (2005). How Can Instantaneous Velocity Fulfil its Causal Role? *Philosophical Review*, 114, 433-468.

Lowe, E. J. (1998). *The Possibility of Metaphysics*. Oxford: Clarendon Press.

Lowe, E. J. (2008). Ontological Dependence. *The Stanford Encyclopedia of Philosophy (Fall 2008 Edition)*, ed. E. N. Zalta, URL: <http://plato.stanford.edu/archives/fall2008/entries/dependence-ontological/>.

Markosian, N. (2010). Time. *The Stanford Encyclopedia of Philosophy (Winter 2010 Edition)*, ed. E. N. Zalta, URL: <http://plato.stanford.edu/archives/win2010/entries/time/>.

McKinnon, N. (2003). Presentism and Consciousness. *Australasian Journal of Philosophy*, 81, 305-323.

Orilia, F. (2012). Dynamic Events and Presentism. *Philosophical Studies*, 160, 407-414.

Owen, G. E. L. (1957-58). Zeno and the Mathematicians. *Proceedings of the Aristotelian Society*, 67, 199-222. (Repr. in Salmon 1970, 139-163; page references are to this repr.).

Paseau, A. (2010). Defining Ontological Basis and the Fundamental Layer. *Philosophical Quarterly*, 60, 169-175.

Penrose, R. (2004). *The Road to Reality*. New York: Knopf.

Russell, B. (1914). *Our Knowledge of the External World*, Allen & Unwin: London.

Salmon, W. (ed.), (1970). *Zeno's Paradoxes*. Indianapolis: Bobbs-Merrill (2nd ed., Indianapolis: Hackett, 2001).

Salmon, W. (1970a). Introduction. Salmon 1970, 5-44.

Salmon, W. (1975). *Space, Time, And Motion*. Encino: Dickenson.

Shoemaker, S. (1969). Time Without Change. *Journal of Philosophy*, 66, 363-381.

Smolin, L. (2000). *Three Roads To Quantum Gravity*. London: Weidenfel and Nicolson.

Smolin, L. (2006). Atoms of Space and Time. *Scientific American*, 16 (1), 82-90.

Sorabji, R. (1983). *Time, Creation and the Continuum*. London Duckworth.

Whitehead, A. N. (1925). *Science and the Modern World.* New York: MacMillan.

Whitehead, A. N. (1929). *Process and Reality*. Cambridge: Cambridge University Press.

Whitrow, G. J. (1980). *The Natural Philosophy of Time*. Oxford: Clarendon Press.

T×W Epistemic Modality[*]

ANDREA IACONA
University of Turin
andrea.iacona@unito.it

ABSTRACT. So far, T × W frames have been employed to provide a semantics for a language of tense logic that includes a modal operator that expresses historical necessity. The operator is defined in terms of quantification over possible courses of events that satisfy a certain constraint, namely, that of being alike up to a given point. However, a modal operator can as well be defined without placing that constraint. This paper outlines a T × W logic where an operator of the latter kind is used to express the epistemic property of definiteness. Section 1 provides the theoretical background. Sections 2 and 3 set out the semantics. Sections 4 and 5 show, drawing on established results, that there is a sound and complete axiomatization of the logic outlined.

1. Introduction

This paper originates from some reflections on future contingents. Among those who have attempted to provide a rigorous account of future contingents, there is a widespread tendency to think that the most appropriate formal semantics for a tensed language involves branching time structures, that is, structures formed by a set of times and a tree-like partial order defined on the set. This inclination is fostered by two assumptions. One is that indeterminism entails *branching*, that is, the conception according to which there is a plurality of possible courses of events that overlap up to a certain point, the present. The other is that an adequate account of the semantic properties of future contingents hinges on the notion of *determinacy*, understood as truth in all possible courses of events. In a branching time structure, overlapping possible courses of events are represented as maximal linearly ordered subsets of times, and determinacy is expressed in terms of truth at a time relative to all possible courses of events that include that time.[1]

However, both assumptions might be rejected. Against the first it may be argued that, at least on a plausible understanding of indeterminism, indeterminism does not entail branching. If determinism is understood as

---
[*] This article is published in *Logic and Philosophy of Science* 2012, vol. 10, n. 1.
[1] The notion of branching time structure goes back to Kripke, see Prior 1967, 27-29.

the claim that for any time, the state of the universe at that time is entailed by the state of the universe at previous times together with the laws of nature, and indeterminism is understood as the negation of that claim, then indeterminism is consistent with a conception according to which possible courses of events do not overlap. Possible futures may be conceived as parts of possible worlds that are wholly distinct, rather than branches that depart from a common trunk.[2]

The second assumption may be questioned in at least two ways. In the first place, it may be argued that any account of future contingents centred on the notion of determinacy neglects a crucial distinction, namely, that between truth and determinate truth. Suppose that the following sentence is uttered now

(1) It will rain

It is at least consistent to claim that (1) may be true even though it is not determinately true, if it is true in the actual course of events but false in some other possible course of events. Secondly, it may be argued that some of the facts that the notion of determinacy is intended to capture in reality are epistemic facts, hence that an account of them in terms of a formal representation of a state of knowledge is preferable to one that depends on unnecessary metaphysical assumptions. Consider the apparent difference between (1) and the following sentence

(2) Either it will rain or it will not rain

This difference can be explained epistemically as follows: now we are not able to tell whether (1) is true because as far as we know (1) is true in some but not in all possible courses of events. By contrast, we can confidently assert (2), as (2) seems to be true in all possible courses of events.[3]

If the two assumptions are rejected, no strong motivation remains for regarding branching time structures as a privileged formal tool to deal with the issue of future contingents. In particular, if the second assumption is rejected on the basis of considerations about the epistemic nature of facts such as that considered, there seems to be no reason to restrict attention to

---

[2] Hoefer considers a definition of indeterminism along these lines, see Hoefer 2010. Lewis argues against branching in Lewis 1986, 206-209. In Iacona 2013 I discuss the claim that indeterminism entail branching.

[3] In Iacona 2013 I argue for the distinction between truth and determinate truth.

metaphysical interpretations of formal semantics. This paper explores one of the alternative routes. The model of time that will be outlined, *the grid model*, belongs to the family of T × W semantics, and the interpretation of it that will be considered is epistemic rather than metaphysical.[4]

2. The grid model

Let Φ be the set of propositional variables. Our language will be the smallest set including Φ that is closed under composition by means of the propositional connectives and the operators G, H and D. Its semantics is based on the following definition.

Definition 1 *Let T and W be sets. A G-frame is a pair $\langle \{T_w, <_w\}_{w \in W}, \approx \rangle$ that satisfies the following conditions.*

1. *For any $w \in W$, $T_w \subseteq T$. For any $w, w' \in W$ such that $w \neq w'$, $T_w \cap T_{w'} = \emptyset$.*

2. *For any $w \in W$, $<_w$ is a linear order on T_w. A relation $<$ on T is defined accordingly: for any $t, t' \in T$, $t < t'$ iff there is a w such that $t, t' \in T_w$ and $t <_w t'$.*

3. *$\approx$ is an equivalence relation on T such that (a) for any $t \in T$ and $w \in W$, there is a unique t' such that $t' \in T_w$ and $t \approx t'$, (b) if $t, t' \in T_w, t'', t''' \in T_{w'}, t \approx t'', t' \approx t'''$ and $t <_w t'$, then $t'' <_{w'} t'''$.*

From now on, Dn will abbreviate 'definition n,' and D$n.m$ will abbreviate 'clause m of definition n.' The numbers of T are called times. The members of W are called worlds. So from D1.1 and D1.2 it turns out that worlds are linearly ordered disjoint sets of times. This means that times are world-relative temporal units, in that each time belongs at most to one world. The relation $\approx$ specified in D1.3, by contrast, expresses the trans-world relation of "being at the same time," so induces a partition of times that is orthogonal to their chaining into worlds. To make this clear, the equivalence classes of times determined by $\approx$ may be called "instants," fol-

---

[4] The original formulation of T × W semantics is given by Thomason in Thomason 1984. The epistemic interpretation that will be considered develops a suggestion that I advanced in Iacona 2009.

lowing the terminology adopted by Belnap, Perloff and Xu in Perloff, Belnap and Xu 2001. In figure 1, worlds are represented as straight vertical lines that run parallel, whereas instants are represented as straight horizontal lines that cut across them: for example, t and t' belong to the same world, while t and t'' belong to the same instant.

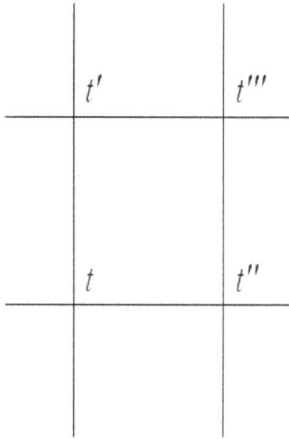

Figure 1: The grid of worlds and instants

Definition 2 *A G-structure is a triple $\langle \{T_w, <_w\}_{w \in W}, \approx, V \rangle$, where $\langle \{T_w, <_w\}_{w \in W}, \approx \rangle$ is a G-frame and V is a function that assigns a truth-value to each formula α for any time t in the following way:*

1. $V_t(\alpha) \in \{1, 0\}$ *for* $\alpha \in \Phi$.

2. $V_t(-\alpha) = 1$ *iff* $V_t(\alpha) = 0$.

3. $V_t(\alpha \supset \beta) = 1$ *iff* $V_t(\alpha) = 0$ *or* $V_t(\beta) = 1$.

4. $V_t(G\alpha) = 1$ *iff for every t' such that $t < t'$, $V_{t'}(\alpha) = 1$.*

5. $V_t(H\alpha) = 1$ *iff for every t' such that $t' < t$, $V_{t'}(\alpha) = 1$.*

6. $V_t(D\alpha) = 1$ *iff for every t' such that $t \approx t'$, $V_{t'}(\alpha) = 1$.*

D2.1-D2.3 are standard. D2.4 and D2.5 specify the meaning of G and H, read as 'henceforth' and 'hitherto.' D2.6 characterizes D, read as 'definitely.' D differs from G and H in a way that is easy to grasp visually. G and H ask you to move along the vertical axis and go up or down on the same world, while D asks you to move along the horizontal axis and go left and right on the same instant.[5]

Other symbols may be added to the language on the basis of D2. $\land$ and $\lor$ depend on $\sim$ and $\supset$ in the usual way. Two operators F and P may be defined in terms of G and H as follows: $F\alpha \equiv {\sim}G{\sim}\alpha$ and $P\alpha \equiv {\sim}H{\sim}\alpha$. Similarly, an operator C may be defined in terms of D as follows: $C\alpha \equiv {\sim}D{\sim}\alpha$. Truth in a structure and validity are defined in the standard way:

Definition 3 α *is true in a G-structure iff for any* t, $V_t(\alpha) = 1$.

Definition 4 α *is valid, that is,* $\vDash \alpha$, *iff* α *is true in all G-structures.*

The semantics outlined is a kind of T × W semantics. In particular, the version of T × W semantics that best suits the present purposes is that provided by Kutschera 1997. Kutschera defines a STW frame as a triple $\langle \{T_w, <_w\}_{w \in W}, \approx, \sim \rangle$, where the first two terms satisfy the conditions specified in D1 and $\sim$ is an equivalence relation that differs from $\approx$ in that it is designed to express historical necessity. So G-frames are nothing but STW frames without that relation.[6]

3. The epistemic interpretation

On the interpretation of the grid model that will be considered, times designate epistemically possible global states of affairs, and worlds are understood as epistemically possible courses of events. The underlying thought is that, for every sentence 'p' such that we are not in a position to know that p, there are at least two worlds: one in which p and one in which it is not the case that p. For example, today we are not in a position to

---

[5] A modal operator defined in terms of unrestricted quantification over worlds, like D, was first considered by Di Maio and Zanardo 1994.

[6] The letter S in STW stands for 'separated,' to distinguish STW frames from standard T × W frames as defined by Thomason 1984. Kutschera 1997 shows that for every T × W structure there is an equivalent STW structure, and vice versa, see p. 243.

know whether it will rain tomorrow. So there are at least two worlds: one in which it rains tomorrow and one in which it doesn't.

The use of the expression 'in a position to know' presupposes that a meaningful distinction can be drawn between knowing that p and being in a position to know that p. Being in a position to know that p, like knowing that p, is factive: if one is in a position to know that p, then it is true that p. But the two states are not exactly the same. While knowing that p entails being in a position to know that p, being in a position to know that p does not entail knowing that p: one may be in a position to know that p without knowing that p, just like one may fail to see something that is in front of one's eyes.[7]

Note that the differences between epistemically possible courses of events are not limited to the future. For example, we are not in a position to know whether the number of cats that slept inside the Colosseum on September 4th 1971 is even or odd. So there are at least two worlds: one in which that number is even, the other in which that number is odd. The same goes for the present. For example, we don't know the exact location of a certain whale that is now swimming in the ocean, so we are not able to discriminate between times that differ as to the location of that whale. The absence of a unique present time is a key feature of the grid model. In figure 1 there is no point that indicates where you are. The reason is that you don't know exactly where you are, in that you don't know which of the two worlds is your world. What allows you to locate yourself on the diagram is a line rather than a point, that is, an instant.

On the epistemic interpretation, D expresses truth in all epistemically possible courses of events. To say that it is definitely the case that p is to say that one is in a position to know that p. For example, the apparent difference between (1) and (2) may be explained in terms of definiteness. Consider figure 1. If p is true at t' but false at t''', then DFp is false at t, while $D(Fp \lor \sim Fp)$ is true at t.

The operator C is construed accordingly. To say that one is in a position to know that p is to say that every epistemically possible course of events is such that p. Therefore, if one is not in a position to know that it is

---

[7] Williamson provides a characterization of the distinction along these lines, see Williamson 2000, 95. In any case, nothing substantial will depend on this distinction. The underlying thought of the epistemic interpretation could equally be rephrased as follows: for every sentence 'p' such that we don't know that p, there are at least two worlds: one in which p and one in which it is not the case that p.

not the case that p, then one is not in a position to exclude that p, that is, some epistemically possible course of events is such that p.

4. Axiomatization

T × W logic has been shown to be complete under two axiomatizations. One is the finite axiomatization adopted by Kutschera 1997, which includes the irreflexivity rule introduced by Gabbay 1981. The other is the infinite axiomatization adopted by Di Maio and Zanardo 1998, which is free from that rule. The system outlined here follows Kutschera, for the completeness proof is simpler with the irreflexivity rule. But a similar system could be constructed in terms of the other axiomatization.

Let S be a system whose axioms include the standard propositional axioms and the following:

A1 $G(\alpha \supset \beta) \supset (G\alpha \supset G\beta)$

A2 $H(\alpha \supset \beta) \supset (H\alpha \supset H\beta)$

A3 $\alpha \supset HF\alpha$

A4 $\alpha \supset GP\alpha$

A5 $G\alpha \supset GG\alpha$

A6 $F\alpha \supset G(F\alpha \vee \alpha \vee P\alpha)$

A7 $P\alpha \supset H(F\alpha \vee \alpha \vee P\alpha)$

A8 $D\alpha \supset \alpha$

A9 $D(\alpha \supset \beta) \supset (D\alpha \supset D\beta)$

A10 $C\alpha \supset DC\alpha$

A11 $DG\alpha \supset GD\alpha$

A12 $DH\alpha \supset HD\alpha$

A13 $FD\alpha \supset DF\alpha$

A14 $PD\alpha \supset DP\alpha$

A1-A7 are standard axioms of linear tense logic. A1-A2 state that distribution holds for G and H. A3-A4 ensure that G and H depend on accessibility relations that are converse to each other. A5 expresses the transitivity of $<$. A6 rules out branching to the future, while A7 rules out branching to the past.

A8-A10 characterize D as a modal operator. A8 expresses a platitude, as it amounts to saying that being in a position to know is factive. A9 is easily justified. If one is in a position to know that if p then q and one is in a position to know that p, then one must be in a position to know that q. For all that is needed to get to the conclusion that q is to apply a valid rule of inference. A10 entails that if for all one knows it could be the case that p, then one is in a position to know that for all one knows it could be the case that p. This is quite plausible. Suppose that one is not in a position to know that it is not the case that p. Then, presumably, the negation of p does not hold in all possible courses of events in virtue of some logical principle, and one is in a position to know that.

A11-A14 state a connection between G, H, F and P on the one hand, and D on the other. According to A11, if it is knowable that from now on it will be the case that p, then from now on it will be knowable that p. This is acceptable if one thinks that the antecedent of the conditional is satisfied only for those truths that hold at any time. For example, it is true at any time in every epistemically possible course of events that if it rains then it rains. Thus, it is knowable that from now on if it rains then it rains. But if so then the consequent is satisfied, that is, from now on it will be knowable that if it rains then it rains. The motivation for A12-A14 is similar.

Let $\vdash$ stand for derivability in S. The rules of inference of S are the following:

R1 If $\vdash \alpha \supset \beta$ and $\vdash \alpha$, then $\vdash \beta$

R2 If $\vdash \alpha$, then $\vdash G\alpha$ and $\vdash H\alpha$

R3 If $\vdash \alpha$, then $\vdash D\alpha$

R4 If ⊢ $D(\sim p \wedge Gp) \supset \alpha$, then ⊢ α, where p is a propositional variable that does not occur in α

R1 is *modus ponens*, R2 is temporal generalization, while R3 amounts to the rule of necessitation. R4 is the version of Gabbay's irreflexivity rule used by Kutschera.

5. Soundness and completeness

S is sound. It is straightforward to verify that A1-A14 are valid and R1-R4 preserve validity. The completeness of S can be proved through the method used by Kutschera for a system called TW. Kutschera defines STW systems as sets of maximal consistent sets of formulas endowed with a relational structure, and shows that STW systems induce STW structures. Thus, in order to prove that TW is complete it suffices to show that for every formula that is not a theorem of TW there is a STW system that includes its negation, for that in turn entails the existence of a STW structure in which the formula is not true at some time. Here a proof will be provided to the effect that STW systems induce G-structures. So the completeness of S will be obtained in the same way, using von Kutschera's result about the existence of a STW system.

Let us grant Kutschera's definition of STW systems. To abbreviate, 'mcs' will stand for 'maximal consistent set of formulas'. The relation R_G is defined as follows: if S and S' are mcss, $SR_G S'$ iff $G(S) \subseteq S'$, where $G(S) = \{\alpha: G\alpha \in S\}$. The relations R_H and R_D are defined in similar way. If S and S' are mcss, $SR_H S'$ iff $H(S) \subseteq S'$, where $H(S) = \{\alpha: H\alpha \in S\}$. If S and S' are mcss, $SR_D S'$ iff $D(S) \subseteq S'$, where $D(S) = \{\alpha: D\alpha \in S\}$. R_G and R_H are transitive, while R_D is an equivalence relation.

Definition 5 *A STW system is a pair* $\langle \{S_t\}_{t \in T}, \{T_w\}_{w \in W} \rangle$ *defined as follows.*

1. *W is a set of indices.*

2. *The sets T_w are disjoint, and T is the union of them.*

3. *For every $t \in T$, S_t is a mcs. Each $t \in T$ has itw own mcs, so if $t \neq t'$ then $S_t \neq S_{t'}$.*

4. For every $w \in W$ and $t \in T_w$, if $F\alpha \in S_t$ then there is a $t' \in T_w$ such that $S_t R_G S_{t'}$, and $\alpha \in S_{t'}$. The same goes for P and R_H. The case of C and R_D is similar, but without the condition that $t' \in T_w$.

5. For every $t, t' \in T_w$, either $S_t = S_{t'}$ or $S_t R_G S_{t'}$ or $S_{t'} R_G S_t$.

6. For every $t \in T$ and for some propositional variable p, $D(\sim p \wedge G_p) \in S_t$.

7. For every $w, w' \in W$ and every $t \in T_w$, there is a $t' \in T_{w'}$ such that $S_t R_D S_{t'}$.

Let it be granted that $t <_w t'$ iff $t, t' \in T_w$ and $S_t R_G S_{t'}$, and that $t \approx t'$ iff $S_t R_D S_{t'}$. Now it will be shown that for every STW system there is a correspondent G-structure.

Theorem 1 *If* $\langle \{S_t\}_{t \in T}, \{T_w\}_{w \in W}\rangle$ *is a STW system, then* $\langle \{T_w, <_w\}_{w \in W}, \approx \rangle$ *is a G-frame.*

Proof. D1.1 follows from D5.2. D1.2 follows from D5.5. To see that D1.3 is satisfied, consider condition (a) first. The existence of t' is entailed by D5.7. The uniqueness of t' is shown as follows. Suppose that $t, t' \in T_w$ and $t \approx t'$. From D5.5 we get that either $S_t = S_{t'}$ or $S_t R_G S_{t'}$ or $S_{t'} R_G S_t$. But the second disjunct cannot hold, because from D5.6 we get that $Gp \in S_t$, hence that $p \in S_{t'}$. Since we also have that $D \sim p \in S_t$, hence that $\sim p \in S_{t'}$, because $t \approx t'$, we get that both $p \in S_{t'}$ and $\sim p \in S_{t'}$, which contradicts D5.3. A similar reasoning shows that the third disjunct cannot hold. Therefore, $S_t = S_{t'}$.

Now consider condition (b). Suppose that $t, t' \in T_w$, $t'', t''' \in T_{w'}$, $S_t R_D S_{t''}, S_{t'} R_D S_{t'''}$ and $S_t R_G S_{t'}$. From D5.5 we get that either $S_{t''} = S_{t'''}$ or $S_{t''} R_G S_{t'''}$ or $S_{t'''} R_G S_{t''}$. But the first disjunct cannot hold. D5.6 entails that $D(\sim p \wedge G_p) \in S_t$, hence that $\sim p \in S_{t''}$. Since we also have that $DGp \in S_t$, A11 entails that $GDp \in S_t$, hence that $Dp \in S_{t'}$ and consequently that $p \in S_{t'''}$. Therefore, the first disjunct contradicts D5.3. The third disjunct leads to a similar conclusion. For D5.6 entails that $D(\sim p \wedge G_p) \in S_t$, hence that $\sim p \in S_{t''}$. Since D5.6, in combination with A5, also entails that $DGGp \in S_t$, by A11 we get that $GDGp \in S_t$, and consequently that

$DGp \in S_{t'}$. Since $S_{t'}R_DS_{t'''}$, it follows that $Gp \in S_{t'''}$. So if it were the case that $S_{t'''}R_GS_{t''}$, we would get that $p \in S_{t''}$. Therefore, $S_{t''}R_GS_{t'''}$.

□

Theorem 2 *For each STW system $\langle \{S_t\}_{t \in T}, \{T_w\}_{w \in W}\rangle$ there is a G-structure $\langle \{T_w, <_w\}_{w \in W}, \approx, V\rangle$ such that, for every α, $V_t(\alpha) = 1$ iff $\alpha \in S_t$.*

Proof. Let $\langle \{S_t\}_{t \in T}, \{T_w\}_{w \in W}\rangle$ be a STW system. Theorem 1 entails that $\langle \{T_w, <_w\}_{w \in W}, \approx\rangle$ is a G-frame. A function V can be defined on the frame in accordance with D2, assuming that, for each α in Φ, $V_t(\alpha) = 1$ iff $\alpha \in S_t$. This way it can be shown by induction on the complexity of α that for every α, $V_t(\alpha) = 1$ iff $\alpha \in S_t$. The case of $\sim \alpha$ and $\alpha \supset \beta$ is trivial. Consider the case of $G\alpha$. Let us assume that $V_t(\alpha) = 1$ iff $\alpha \in S_t$, and suppose that $V_t(G\alpha) = 1$. From D2.4 we get that for t' such that $t < t'$, $V_{t'}(\alpha) = 1$. Since $t < t'$ iff $S_tR_GS_{t'}$, $G\alpha \in S_t$ if $\alpha \in S_{t'}$. So $G\alpha \in S_t$. Now suppose that $G\alpha \in S_t$. Since $t < t'$ iff $S_tR_GS_{t'}$, for every t' such that $t < t'$ we get that $\alpha \in S_{t'}$, hence that $V_{t'}(\alpha) = 1$. So D2.4 entails that $V_t(G\alpha) = 1$. The case of $H\alpha$ and $D\alpha$ is similar. Therefore, $\langle \{T_w, <_w\}_{w \in W}, \approx, V\rangle$ is a G-structure such that, for every α, $V_t(\alpha) = 1$ iff $\alpha \in S_t$.

□

Kutschera proves that if a formula is not a theorem of TW, there is a STW system $\langle \{S_t\}_{t \in T}, \{T_w\}_{w \in W}\rangle$ such that for some $t \in T$, the negation of the formula belongs to S_t. A similar result holds for S, that is,

Theorem 3 *If it is not the case that $\vdash \alpha$, then there is a STW system $\langle \{S_t\}_{t \in T}, \{T_w\}_{w \in W}\rangle$ such that for some $t \in T$, $\sim \alpha \in S_t$.*

Proof. Kutschera (1997, 246-247) shows how theorem 3 can be proved in two steps. First, Gabbay's irreflexivity lemma can be used to show that if it is not the case that $\vdash \alpha$, then there is a set $\{S_t\}_{t \in T}$ that satisfies certain conditions and a t_0 such that $\sim \alpha \in S_{t_0}$ (theorem 4.1). S is like TW in this respect, as it includes the rule R4, which is required by the proof. Second, a STW system can be constructed from $\{S_t\}_{t \in T}$ by defining a set of $\{T_w\}_{w \in W}$

(theorem 4.2). Again, S is like TW in this respect, as it includes the axioms used in the proof.

□

Theorem 4 *If* ⊨ α *then* ⊢ α.

Proof. From theorems 2 and 3 it follows that if is not the case that ⊢ α, then there is a G-structure such that $V_t(\alpha) = 0$ at some t.

□

REFERENCES

Di Maio, C., and Zanardo, A. (1994). Synchronized Histories in Prior-Thomason Representation of Branching Time. *Proceedings of the First International Conference on Temporal Logic*, eds. D. Gabbay and H. Ohlbach, Springer, 265-282.

Di Maio, C., and Zanardo, A. (1998). A Gabbay-Rule Free Axiomatization of T × W Validity. *Journal of Philosophical Logic*, 27, 435-487.

Gabbay, D. (1981). An Irreflexivity Lemma with Applications to Axiomatization of Conditions on Tense Frames. *Aspects of Philosophical Logic*, ed. U. Mönnich, Reidel.

Hoefer, C. (2010). Causal Determinism. *The Stanford Encyclopedia of Philosophy* (Spring 2010 Edition), ed. E. N. Zalta, URL: <http://plato.stanford.edu/archives/spr2010/entries/determinism-causal>.

Iacona, A. (2009). Commentary on R. Thomason 'Combinations of Tense and Modality'. *Humana.Mente*, 8, 185-190.

Iacona, A. (2013). Timeless Truth. *Around the Tree: Semantic and Metaphysical Issues Concerning Branching and the Open Future*, eds. F. Correia and A. Iacona, Springer, 29-45.

Lewis, D. (1986). *On the Plurality of Worlds*. Oxford: Blackwell.

Perloff, M., Belnap, N., and Xu, M. (2001). *Facing the Future*. Oxford: Oxford University Press.

Prior, A. N. (1967). *Past, Present and Future*. Oxford: Clarendon Press.

Thomason, R. H. (1984). Combinations of Tense and Modality. *Handbook of Philosophical Logic*, Vol. 2, eds. D. Gabbay and G. Guenthner, Reidel, 135-165.

von Kutschera, F. (1997). T × W Completeness. *Journal of Philosophical Logic*, 26, 241-250.

Williamson, T. (2000). *Knowledge and Its Limits*. Oxford: Oxford University Press.

TOWARDS A THEORY OF MULTIDIMENSIONAL TIME TRAVEL

DOMENICO MANCUSO
Università di Urbino
philonous@libero.it

ABSTRACT. This paper outlines a non-standard temporal model which might be compatible with time travel: on one hand, because it averts the causal loops arising in a linear chronology; on the other, because it incorporates certain formal requirements which are implicit in the way time travel is usually conceived. The most important such requirement stipulates that each agent have permanent access to a "private past" which no one else can modify. As it turns out, the minimal model satisfying the given conditions involves $n+1$ dimensions for n independent travellers, with adequate geometric constraints on each agent's local displacements.

Conjectures to the effect that time could have more than one dimension are progressively coming to be regarded as scientifically respectable views (although very marginal ones). In physics, the rationale behind such proposals has usually to do with efforts aimed at unifying general relativity and quantum mechanics; the most knowledgeable advocate of this approach is probably Izthak Bars, who first put forth the idea of two-dimensional time (and four-dimensional space: see Bars, Deliduman and Andreev 1998) and has since then consistently elaborated on that insight, through more than forty published articles; other physicists argue, on different grounds, for a model with six coordinates, three spatial and three temporal (see for example Chen 1999 and 2005, Sparling 2006).

From a philosophical standpoint, motivations behind many-dimensional approaches are more varied. While Steven Weinstein (2008) retains a close connection with physical theories, Robert Vallée—whose primary interest is in cybernetics—has used the idea in the context of an analysis on memory and perception (1991). By contrast, Nathan Oaklander's contribution is genuinely metaphysical: in a 1983 paper, he brilliantly argued that the tensed view of time (the A-series) can only be formalized through a multiplication *ad infinitum* of temporal dimensions, which is perfectly parallel to Mc Taggart's well-known infinite regress of temporal attributions.

It is doubtful whether a comprehensive analysis of such heterogeneous literature (and other similar examples) could reveal any common ground, besides asuperficial similarity; in any case, it is out of the scope of the present paper. What I intend to do is, somehow more modestly, focus on a specific aspect of higher-dimensional models: namely, their possible application to the philosophical discussion on time travel, as a conceptual tool for sidestepping the logical problems that arise in a standard topology.

In this restricted sense, previous accounts of many-dimensional time may be reduced to three. The earliest reference, to my knowledge, is to be found in C.D. Broad, who framed the general idea in a section of his 1937 essay on the implications of foreknowledge (pp. 200-4). Jack Meiland (1974) took a step forward and laid out the basic formal machinery for two-dimensional time travel. Finally, Goddu (2003) followed in Meiland's trail, emphasizing the idea of *changes* to the past and engaging related metaphysical issues such as personal identity and analogies with branching time.

Meiland's and Goddu's works, in my view, are only a sketch of what could be a comprehensive theory of multidimensional time travel. This essay is aimed at laying out the groundwork for such a theory, first by supplying the existing 2-D model with an interpretation in terms of subjective and objective time, and secondly—and most importantly—by generalizing to an arbitrary number of dimensions. Although it is not required to avert logical paradoxes, the second step is justified as a response to certain additional assumptions that are intuitively associated with time travel stories; these assumptions, and their mutual connection, will emerge in the course of the paper.

The conclusion will leave open a number of substantialquestions, and pave the way to further generalizations, to be carried out by lifting some formal restrictions that I am going to impose here for the sake of simplicity. Thus, rather than as an accomplished work (which they partly are), the pages that follow should be seen as an open-ended project that will hopefully merge into a more systematic study.

1. Extra dimensions—what for?

There is one question that could very easily be raised before we even begin to construct the mathematical model: what is the whole point of it? Why should we introduce multiple dimensions at all?

As a metaphysician educated in pure mathematics, my favourite reply would be that the model may be studied simply for the sake of its formal structure, just as we could do—at least in the first instance—with a system of logic.

I am perfectly aware, on the other hand, that few people will find such a response satisfying, especially if the question is supported by an appeal to the principle of simplicity: dimensions, as Ockham would have it, *non sunt multiplicanda* if a single one can do the job. The complex model, no matter how formally interesting, may only be brought in if we can point at precise shortcomings of the simpler explanation.

The main such shortcoming is obviously the logical inconsistency of standard (i.e. one-dimensional) time travel, which I will briefly illustrate in this section. Subsequently, while developing the multi-dimensional framework, I will point at the fact that a linear temporal topology fails to mirror some intuitions which are deeply embedded in our conception of time travel, and I will discuss to what extent this could constitute an independent justification for introducing the new model.

Claiming that one-dimensional time travel is paradoxical means rejecting at once two theses which are quite fashionable in the philosophical community, although by no means undisputed:
1) time travel must be admitted in principle because it is compatible with our best physical theories, i.e. general relativity and quantum mechanics—or at least, it may not be conclusively shown to be incompatible;
2) even from an *a priori* standpoint, travelling backwards in linear time implies at worst extremely unlikely scenarios, but no true logical paradox.

Needless to say, I may not properly address these contentions here, since either of them would require far more than an article-length essay for a reasonably complete account. What I shall do, with no pretension to being exhaustive, is to hint at some arguments for refuting both views; essentially, though, the reader is asked to accept the incoherency of standard time travel as a premise for discussing its multi-dimensional counterpart—to accept it, if not as an ontological commitment, at least as a conditional assumption.

Incidentally, a similar ontological disengagement may be entertained towards multi-dimensional time itself. The temporal framework I will outline need not be thought of as the *real* structure of time; in fact, I shall lay no such claim at all: the function of the multidimensional model, within this paper as well as in any future development, is simply that of a mathematical tool.

Let us now revert to the first thesis mentioned above: is it true that physical research has said the last word on the logical coherency of time travel?

As a general rule, in a philosophical debate, arguments of a different kind may have very different weights: in particular, those of a logical nature—that is, concerning the possibility of formulating and representing a notion—should always have seniority rights on empirical considerations, including those appealing to the most rigorous scientific theories. *A priori* arguments constitute, so to speak, a *conditio sine qua non*: if some reality may not be rationally conceived, it is pointless to ask whether it can be admitted in a given empirical context; the latter question only becomes significant, and perhaps decisive, once that reality has been assumed to be possible in principle.

Physical explanations are fallible by nature, even those which seem unassailable at present; better still, they are all awaiting their disavowal, and therefore can only ground short-lived and transient philosophical theories. This is all the more true for the issue of time travel, where physical arguments do not stem directly from general relativity or quantum mechanics, but rather from some peculiar interpretations, leading to hypotheses which are controversial among physicists themselves: think, for example, of Novikov's "self-consistency principle" as opposed to Hawking's "chronology protection conjecture,"[1] or else of the occasional experiments purporting to prove reverse causation, which have all been turned down by the scientific community since alternative explanations were apparently available. Arguments of this kind, which have no general consensus within their own discipline, could hardly bear the burden of a thesis such as the possibility of time travel, whose implications are far-reaching and involve deep philosophical problems like free will, personal identity, or the continuity of consciousness.

Coming to the second thesis, i.e. to the conceptual arguments, time travel has an obvious *prima facie* inconsistency: a traveller might alter the past in

---

[1] The Novikov principle permits time travel under strong constraints, that is, the traveller is allowed to alter past events but may not change in any way the phenomenal evidence they have left in the present; how exactly such a constraint can be built into the universe is obviously quite unclear. Hawking's conjecture, on the other hand, forbids time loops altogether.

such a way that his own journey becomes impossible; this is sometimes known as the "Grandfather paradox," because of David Lewis' (1976) well-known example of a young man killing his own ancestor.

The standard reply to this difficulty consists of imposing appropriate constraints on the traveller's behaviour, such as a series of coincidences preventing him from performing the paradoxical act. ("Perhaps some noise distracts him at the last moment, perhaps he misses despite all his target practice, perhaps his nerve fails, perhaps he even feels a pang of unaccustomed mercy"[2]). Yet, even supposing that such a strategy might be successful in a single instance of causal loop, as in Lewis' story, it can hardly be accepted as a general solution, as it would amount to nothing else than a systematic use of *ad hoc* hypotheses.[3]

In fact, cases of travellers killing younger selves or progenitors, as well as other self-refuting scenarios, are just the tip of the iceberg, for they spectacularly bring out the paradoxical nature of time travel; yet, such nature is latent in the background even in situations where changes to the past are limited or "harmless." Apart from whatever contradictions time travel stories may trigger, there is a problem with the very fact that they are *stories*, i.e. with their *dynamic* character: an event E_1, which is *initially* part of the voyager's past, is *subsequently* replaced by an alternative event E_2. This is clearly illogical if we assume the ordinary linear conception of time: each instant is associated to a single event, therefore, if the "real" past event is E_2, it should have been so *since the beginning* of the story.

Thus, most philosophical advocates of time travel part ways with the fictional literature and relinquish the dynamic approach, envisaging instead a *static* model, i.e. a spatio-temporal pattern featuring some local closed loops alongside with open curves. Despite its apparent coherence, such a model entails some serious conceptual problems.

First of all, if the life of an individual comprises one or more circular stages, there appears to be a discontinuity in his personal history: since a truly circular process repeats itself indefinitely, both in the past and in the future, nothing—except an irrational "leap" from another life-stage— explains how someone can get caught into the loop, or get out of it.

Furthermore, even if we set aside this basic incoherence, a life history so structured would involve the sudden appearance of information

---

[2] Lewis 1976, 150.
[3] The standard reference for this kind of argument is Horwich 1987, chapter 7.

stemming from nowhere. This is sometimes referred to as the "writer's paradox": the author of a successful novel sends back the manuscript to his younger self in a time capsule, which is evidently how he originally came to write the book. Who conceived the story in the first place? Apparently, no one did.[4]

Finally, in order to allow static time travel, we need to commit ourselves to a tenseless theory of time. A universe encompassing local loops is perfectly consistent if viewed "from above," with all the paths already traced, and the number and shape of closed curves fixed forever; yet, from the perspective of an agent who is in the midst of a loop, the universe in question is just a provisional arrangement of events which may be changed through a sheer act of will—either by closing the loop in a different way or by turning it into an open curve, or even by setting up a new loop in a region of space-time which hitherto had none. If, on the other hand, we want to rule out *ex hypothesi* any such alteration, a different metaphysical commitment is required: we must assume *compatibilism* in order to reconcile our agent's free will with the fact that the future is written in advance—and perhaps, a conditional analysis *à la* Moore (1912, chapter 6) in order to salvage the intuition that the agent *can* make alternative choices.[5]

Indeed, even Moore unconditionalism will not suffice, since the future is not simply predetermined: it is *known* by the agent—or, at least, he can be made to know it, by transmitting relevant information around the loop. Therefore, the static time travel scenario is analogous to the case of *prediction*, which philosophers have overwhelmingly discarded as inconsistent with free will—not to be confused with the problem of future contingents (which concerns the weaker claim of prior *truth* of certain propositions) and its theological equivalent, divine foreknowledge: unlike prediction, these are controversial, for agent awareness is not assumed.

---

[4] Tentative justifications of the paradox may be found in Levin 1980 and Hanley 2004, 135-139.

[5] Lewis (1976) underscores the ambiguity of the verb *to can* as a key for a compatibilist solution, although he does not endorse conditionalism, nor any other single interpretation. For a sharp criticism of the compatibilist view, see William Grey (1999, 64-5), who connects directly time travel with *fatalism* on free will.

2. Two chronologies

The paradoxes I have briefly illustrated—as well as others that I shall not dwell on—could all be traced back to a single root, *viz.* a discrepancy between two chronologies: an *objective* one, which is the historical succession of events with their dates and times, and a *subjective* one which corresponds to the unfolding of the voyager's conscious experience. The linear topology conflates these chronologies; this is irrelevant in normal circumstances, since they perfectly coincide, but if an agent should move backwards in time, a dissociation would occur: while the objective timeline "folds up" in order to reach a point in the past, the sequence of events in the agent's mind keeps moving forward, because the time of consciousness is by necessity linear.

To construct a temporal model that is hospitable to time travel, we must replace outright identity of objective and subjective time with a looser relationship, namely, a functional dependence of the former on the latter: this will allow agents to revert to an earlier date without violating the continuity of consciousness. The simplest representation of such dependence is an ordinary Cartesian plane, where the x and y stand respectively for subjective and objective time.[6] In this framework, the life of an individual can be described through a continuous function which is generally non-invertible: while I can only place myself in one moment of history at every moment of my inner chronology, it is possible in principle to "visit" the same date in two consecutive stages of life. This is precisely what happens with time travel: instead of a point moving backwards and forwards along a straight path, and passing twice on the same spot (as in the standard model), we have a curve with two points sharing the same y. The derivative $f'(x)$ measures the "velocity" of the traveller; backward journeys take place in those intervals where the function is decreasing, while "ordinary" existence in the present corresponds to a 45° positive angle.

---

[6] In both Meiland's and Goddu's models, the ordinate is the non-reversible one, i.e. subjective time (although neither author mentions the subjective-objective dichotomy); I have switched coordinates to avoid inconsistency with the conventions of analytic geometry.

Essentially, then, inner time plays the role of time proper, whereas the objective chronology is construed by analogy with space,[7] given that functional dependence is the one feature that formally characterizes space with respect to time. The analogy is confirmed by the reference to typically spatial notions such as velocity or direction, which may be applied to historical time once a proper metric has been introduced.

The possibility of spatializing time (at least under a certain description) is an important corollary of the two-dimensional model. Spatial metaphors are deeply entrenched in the way we commonly think about time, as is witnessed by the very use of a word such as "travel"; dissociating subjective and objective chronologies allows to license these analogies, which would be inappropriate in a one-dimensional framework.[8] Thus, for example, we may account for the pervasive image of a "flow of time," which has been the object of sustained attention in analytic philosophy, especially since J. J. Smart (1949, 484) pointed at the need for a "second time-scale with respect to which the flow of events along the first time-dimension is measured": according to the model I have devised, Smart's meta-time could be simply the non-reversible time of consciousness.

Likewise, the second dimension vindicates a seemingly paradoxical notion which is implicit in most time travel stories: that of a "simultaneity" between the traveller's present and the age he is visiting—as for example when a character who has remained "at home" wonders what the hero might be doing "right now." This one-to-one correspondence of events may be interpreted as identity of the abscissae, that is, vertical alignment of points in the Cartesian plane.[9]

---

[7] The relevance of this correlation to time travel in general has been justly emphasized by William Grey (1999, 68): "The whole notion of time travel rests on an unfortunate spatial analogy."

[8] Clearly, setting up a formal apparatus where time behaves like a spatial variable does not solve the problem of whether the analogy is ontologically grounded: the answer to this will depend on whether the model correctly represents the structure of time.

[9] Meiland (1974, sections 2, 7, 8) discusses at length the problem of simultaneity within his own model.

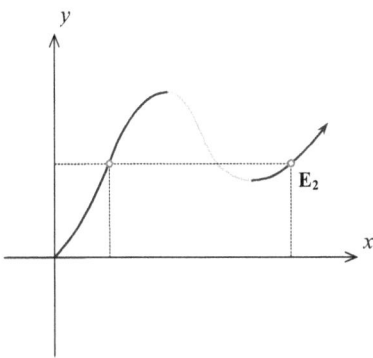 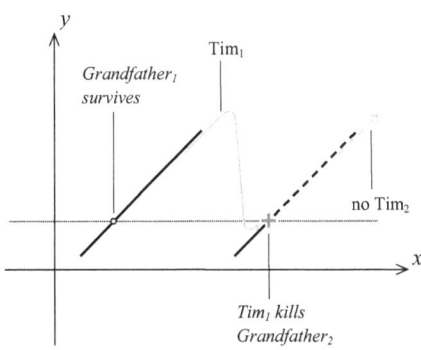

Picture 1: objective and subjective time. The subject lives event E twice; the light section of the path corresponds to an interval where $f(x)$ is decreasing, i.e. objective time runs backwards.

Picture 2: solution of the Grandfatherparadox in 2-D time.

The most important consequence of spatializing historical time, however, concerns the dynamic character of time travel, which I have previously criticized, and which may now be explained away:[10] whenever an agent interferes with a past event, in the 2-D model, it is legitimate to say that the event occurs *first* in a manner and *then* in another one. In linear time, this would amount to a contradiction, since events—unlike objects—*already possess* a specific temporal location, and therefore can neither persist nor change, i.e. display different contents at different moments: thus, for instance, Gandhi's assassination in New Delhi does not extend in time, as opposed to the town, or the gun that shot the bullet, or the victim himself. In two dimensions, the murder does have a location relative to objective time (the *y*-axis) but may still extend along the *x*-axis (while New Delhi and the Mahatma extend along *both* temporal directions). It follows that we can in principle get back to the same event at different moments of our personal history, just as we can do with points or regions of space: in that case, our trajectory in two-dimensional time would be similar to that ex-

---

[10] As I have argued, many philosophers favour the static interpretation, despite its potential conflict with free will, precisely because they regard the dynamic one as implausible; thus, searching for a temporal framework that legitimizes the latter amounts to addressing the conceptual problems of time travel as a whole.

emplified in picture 1, E_1 and E_2 being respectively the original event and its amended "replay"—for instance, Gandhi's shooting as we know it and a failed attempt at his life in the same setting, averted by a time traveller who had witnessed the murder as a child.

How does the bidimensional model deal with a traditional argument such as the Grandfather's paradox?[11] The answer is illustrated in picture 2: Tim_1, who shares our position in 2-D time (and whose existence depends on $Grandfather_1$), gets aboard a time machine and travels to the past (with respect to y); once there, he kills Grandfather—not the one he knew as a child, whom he could only reach through an impossible reversal of subjective time, but $Grandfather_2$, a "duplicate" whose life follows a different path on the temporal plane. Although the murder obliterates any possibility that a Tim_2 be born in the future, it has no *prima facie* consequences on Tim_1's life.

Such consistency is not philosophically cost-free. One obvious shortcoming is that the model seems to commit to the existence of mirror images of the actual world, populated by perfect doubles of the agents we are acquainted with; this leads to a problem of partial identity among individuals, just as we have seen for events. Moreover, the ontological status of an agent's counterparts needs to be determined—are they subjects in their own right, or are they mere copies taken after a blueprint living in the present?

Despite these conceptual problems—as well as others of the same kind—the two-dimensional framework is a clear step forward with respect to linear time: fundamental as they are, concerns about identity do not *a priori* undermine the possibility of discussing time travel, as logical paradoxes do in one dimension. In fact, assessing these concerns, and deciding whether they may be addressed without surrendering basic intuitions on personal identity, is an important thread for further investigation into the model I am sketching in these pages.

---

[11]Goddu 2003 devotes several pages (18; 22 ff.) to the solution of paradoxes, first constructing two examples of direct self-destruction, and then showing how each can be accounted for in a different way within a two-dimensional setting.

2.1. Time loops

So far, the temporal topology I have put forth does not significantly depart from Meiland's or Goddu's, except for a twofold interpretation: philosophically, in terms of objective and subjective time,[12] and mathematically, through the notion of real function which has brought about the coordinate reversal.

This solution is certainly successful in the minimal sense of making time travel a logically consistent hypothesis; nonetheless, it does not fully do justice to certain requirements we intuitively attach to this hypothesis, and consequently, to the philosophical motivations that are behind it. My contention is that such motivations can only be accounted for by indefinitely extending the number of dimensions; in order to see why, it will be useful to make a short detour into fictional literature.

In the narrative of time travel, there exists a subgenre dealing with so-called "time loops," or "temporal loops:" that is, situations where one character (rarely two or more) is forced into going repeatedly through the same interval of time, although he can give events a different turn by reacting in novel ways; on each repetition, everything is reset apart from the memories of the protagonist. Eventually, as a rule, the loop will be broken and a linear chronology will be restored.

This kind of stories was popularized in the nineties by the movie *Groundhog Day*, where a TV weatherman (Bill Murray) keeps waking up on February 2nd in a small village in Pennsylvania, just before a local propitiatory ceremony that he was supposed to cover. Another (deservedly) famous example is *The Butterfly Effect*, where the flashbacks are irregular and unpredictable, they cover several years each, and they leave the protagonist conscious for only a few moments—enough to make a crucial decision—before taking him back to his early twenties to cope with the consequences of that decision. Behind this "mainstream" fiction, there are scores of other variations on the theme, featured in books, movies, TV series and cartoons, including a short novel written as early as 1941[13]—along

---

[12] Lewis 1976 theorizes a distinction between "personal" time, as he calls it, and objective one, but rejects the second dimension; as a result, to preserve consistency, he is forced to discard the subjective chronology, claiming that it does not even qualify as time *stricto sensu*.

[13] Cf. Jameson 1941; other examples include Ballard 1956 and episodes from series such as *Star Trek*, *Stargate*, or *The Twilight Zone*.

with a neglected film called *12:01*, which very probably inspired *Groundhog Day* but was left uncredited.

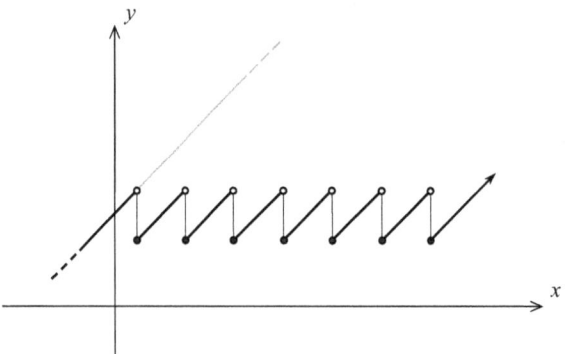

Picture 3: a temporal loop involving repetitions of equal intervals, interpreted within the two-dimensional framework; the light diagonal line indicates the path that the agent would have followed without the loop.

Despite their formal differences, all these stories fall within the bidimensional model as special cases[14] (in fact, a detailed comparative study would probably be very helpful for illustrating the model and its implications). What happens, for each of them, is that the function describing an agent's life path has a discontinuity, dropping from a point (x, y_1) to a point (x, y_2), with $y_1 > y_2$, and then "works its way back" to y_1—or at any rate increases steadily—before being again discontinued:[15] picture 3 gives an example of the resulting graph in the most common narrative pattern.

Are there any specific features of time loops that mark them out with respect to general time travel in two dimensions? Looking at the literature,

---

[14] A notable exception is Ken Grimwood's *Replay*, where a man re-enacts shorter and shorter periods preceding his death: the underlying temporal topology might be described as a many-dimensional extension of ordinary time, yet not of the same kind as the one I discuss in the pages that follow.

[15] If a break in objective time poses some conceptual problems, the continuity of the function can be preserved by postulating a rapid decrease (with $f'(x) \ll 0$) within intervals of subjective time whose length has a fixed upper bound ε.

two plausible candidates are the (momentary) impossibility of overstepping a given value of *y*, and the fact that agents do not control their temporal displacements; however, scenarios with one or both constraints lifted are easily imaginable.

A further difference, and a more substantial one, is the fact that an agent's path in a time loop appears to be broken—as if the traveller's consciousness "leaped" instantly into the body of one of his counterparts. Yet, even this peculiarity might be suppressed if we construed the path as a continuous curve of the kind shown in picture 1. As a corollary, then, we would have to admit the possibility for a traveller to get directly in touch with any of his counterparts (whose paths are independent of his own, and therefore will not systematically swerve in order to avoid crossing it). Such "close encounters" are certainly a counterintuitive hypothesis, but not a strictly paradoxical one, as long as the *alter ego* I run across is ontologically distinct from myself, i.e. someone I can only get to know from a third-person perspective: this, however, is guaranteed by the assumption that consciousness cannot "jump" between two curves. Even a limiting case such as self-destruction would not be essentially different from a murder, perhaps that of a twin brother or sister.

A comparison of theoretical arguments in support of each conception of time loops—the "continuous" model and the "broken" one—is beyond the purpose of this work. I will only mention two reasons that incline me towards the former: first, interruptions of the path pose an obvious problem with the unity of consciousness (which fictional literature usually circumvents through a narrative device, such as putting the character to sleep), and secondly, the continuous paradigm may be easily encompassed as a special case in our general theory of multidimensional time. At any rate, internal consistency of an agent's history is assured in both cases: broken or not, the path he describes may only extend towards the right, and the same holds for the consequences of his actions; even a destruction of duplicate selves—be it experienced as homicide or suicide—could never modify the agent's *past* experiences.

2.2. *The privacy requirement*

When a narrative scheme is exploited as frequently as is the case for time loops, it might be nothing but a literary trend, but it might also descend from some deeper motivations that deserve to be investigated.

To see if some philosophical lesson may be learnt from so frivolous a subject, let us first ask ourselves what is the precise narrative function of temporal loops. The answer, I believe, is that they sustain the teleological perspective (sometimes tinged with moralism) of the stories that revolve around them: typically, the protagonist is caught into the loop for a purpose, such as amending some reprehensible behaviour or grasping an important truth. This is possible because the past is open to repetition and therefore *perfectible*: it can be "rehearsed" until it conforms to a certain standard. Furthermore, this is not a peculiarity of the time loop as such, but a general property of time travel in two dimensions: if we suppress the first two constraints mentioned above—the upper bound for y and the lack of agent control[16]—the only changes will be that the criteria of perfection will be set directly by the agent, rather than by some invisible *deus ex machina*, and that they will apply to an even greater extent, since every spell of one's past life can be chosen for being indefinitely amended—or, using Goddu's (2003, 21 ff.) illuminating metaphor, rolled back and overwritten like a videotape.

Let us now take a step forward and suppose that the agents who can do the overwriting are two or more: could we still claim that the past is perfectible? Certainly not, because the standards of perfection are heterogeneous, and cannot be combined (not to mention the possible intervention of a "villain" who would disrupt the whole teleological fabric of the world). Thus, another condition would have to be assumed in order to guarantee perfectibility: I shall call it the *privacy requirement* (or *privacy condition*). Briefly, it means that, whenever a time traveller alters a past event, he must have the opportunity of doing so in such a way that, once he has gone back to the present, no one can undo the changes; this implies that every agent should have exclusive access to a portion of time including at least one copy of each past event.

The two requirements of privacy and perfectibility are neither arbitrary nor fortuitous: each of them responds to a profound intellectual need which lies at the root of the literary and philosophical interest for time travel—respectively, the desire that the past be preserved somewhere in its integrity, and the possibility of amending personal mistakes.[17] These aspi-

---

[16]The third constraint, *viz.* discontinuity of the path, is wholly irrelevant to the teleological interpretation of the model, for it only concerns the *form* of the repetitions, regardless of their ultimate aims.

[17] The two issues are clearly connected. Formally, they could be described as concerns about the existence of the past in our timeline and about that of alternate timelines, i.e.

rations are in turn the flip side of two dramatic aspects of the human condition, which have driven so much philosophical speculation: the ontological insubstantiality of the present—which subsides into nothingness as soon as it ceases to be actual—and the non-inclusiveness of human choices, which always imply a permanent loss of a possibility of existence. The latter has been extensively discussed by Kierkegaard and the existentialists, as well as exorcized, in different ways, by philosophers of necessity such as the Stoics, Spinoza or Hegel; the former has drawn the attention of analytic philosophers, as is proven by the ongoing debate between presentists and eternalists on the status of the past (and the future).

It goes without saying that I shall not embark into such a far-reaching historical analysis—which the authors I have mentioned would only serve to introduce. The reason why I have hinted at these issues is simply to underscore the philosophical implications of a seemingly trifling subject like time travel—what's more, multidimensional one.

In the pages that follow, I will attempt to develop a description of time travel which, besides being non-contradictory (as is already the case with the two-dimensional solution), may do full justice to the intuitions on perfectibility and privacy. Before doing so, however, I must introduce some more formal apparatus for our 2-D model.

3. *Multiple temporal agents*

In order to assess compliance with the privacy condition, we need an account of how events propagate in the Cartesian plane; this can be easily deduced from the separation of chronologies. Since the ordinate represents objective time, spontaneous changes in physical reality (i.e. those which do not involve voluntary agency) presumably take place along the vertical direction. On the contrary, moving horizontally *per se* does not imply any change: quite simply, a "snapshot" of every event in my past will be carried along on the right of its temporal location (which is a point on a curve in $\mathbf{R}^2$); should I produce some change by time-travelling, the process will be replicated with the modified state of affairs.

From now on, I will call *scattering lines* the oriented half-lines projecting an event from a point $P(x,y)$ to the future, along either dimension;

---

other possible worlds; in the language of modal logic, both may be construed in terms of possible worlds.

moreover, I will refer to horizontal lines as *static* and vertical ones as *dynamic*.

Thus, we have identified two distinct relations holding between moments—or "points"—of two-dimensional time: *identity* in the direction of the x-axis, and deterministic *causation* along the y's—or whatever other basic relation your favourite ontology provides you with. But this is not the whole picture, since another kind of change is brought by voluntary agency, which deploys itself along an individual's life path. Once again, as

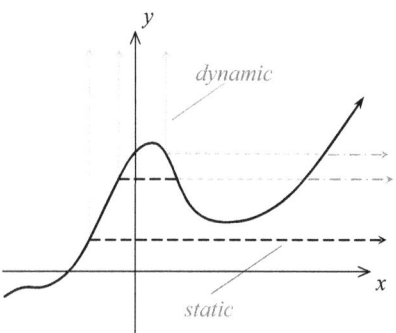

Picture 4: scattering lines

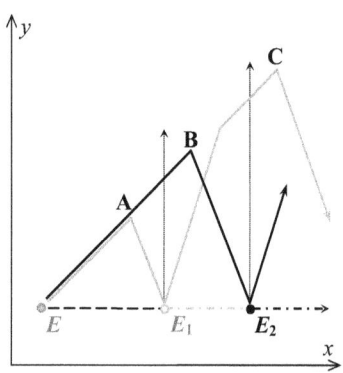

Picture 5: interference with time travel

in the one-dimensional topology, our attitude to time travel appears to have a bearing on the free will problem, although not in the same way: while in that case there seemed to be a conflict, here freedom may not be questioned, because it is essential to the model; were it denied or declared illusory, only the vertical dimension would be significant, since all lines parallel to the y-axis would have the same content. The second dimension may then legitimately be interpreted as the locus of free agency, as opposed to natural causation.

At the same time, the model puts some constraints on how freedom is conceived. To begin with, it implies that an agent does not create events out of thin air, but rather intervenes on a pre-existing state of affairs, transmitted through scattering lines, which could be dubbed a "hypothetical" or "default" reality. This idea of a limited freedom, which is exerted by shaping a sort of raw material, may be intuitively appealing but requires an explanation of *how* exactly the intervention takes place, and how the hypothetical action relates to the accomplished one.

A further problem concerns the extent of the process that produces the default reality. In the movie *Butterfly Effect*, the protagonist repeatedly changes episodes of his past, each time taking a "leap" to the amended present. Since these narrative blackouts are instantaneous (from the standpoint of subjective time), they may be pictured as vertical upward displacements, hence discontinuities of the function, which imply no exercise of free will. And yet, upon returning to the present, the protagonist has to face the consequences of acts and choices that he has purportedly performed, even though he actually wasn't there. This means that "default" deterministic processes implicate the will of agents as well as external objects. One possible explanation could be that they invest a level of subjectivity below consciousness—driving the will like an automatic pilot—and that free agency occurs subsequently over and above that level. Alternatively, we could attribute the determinism to past alter egos of the agent rather than to himself.

At any rate, none of these considerations should be seen as binding: the intuitions that underpin the debate on freedom are far stronger, and far more crucial in defining who we are, than any conjecture on time travel or temporal topology. In fact, conformity between the multi-dimensional model and these intuitions could be a significant test for assessing the plausibility of the model itself.

We are now in a position to take up the problem of interferences.

Suppose I have the capacity to travel in time, but I am not the only one who can do so. As the figure shows, I might decide to set out from the diagonal line at point A, change the outcome of a past event E, and then immediately return to the present; at this stage, someone else could already perceive the effects of the vertical scattering produced by the new event E_1 (see vertical line), and decide in turn to voyage to the past in order to replace E_1 with a further version of the same event. As a result, if I ever found myself once again on the same horizontal line, I would no longer recognize the situation I had set up.

This is a plain disproval of the privacy condition: instead of the personal past we were looking for, all we can get here is a *social* past, which everyone contributes to building. What is worse, other agents might bring their own changes to the original event in such a way that I could not be able to detect the precise origin of the interference, i.e. the objective time coordinate and/or the spatial coordinates of the point where the change occurred; as a consequence, I would presumably also be incapable of neutralizing the effects of their move if I wanted to do so.

The two-dimensional model is designed for time-travelling in a world with a single free agent—whereas everyone else is supposedly obliged to follow the main diagonal; as it stands, it does not support any generalization to cases with two or more independent travellers. On the other hand, if we expand the scheme by adding further dimensions, we can not only account for the privacy condition in these cases, but even grade the level of its implementation.

3.1. Degrees of privacy

First of all, how many dimensions should we introduce? The answer is quite immediate: in order to accommodate m independent time travellers, we must allow for exactly as many variables, plus the usual objective coordinate. The most appropriate model is therefore the real Euclidean space, $\mathbf{R}^n$, where $n = m+1$; in case our world has an infinity of subjects, the geometric representation will accordingly be infinite-dimensional.

Just as in the 2-D model, the temporal path of every individual will be an oriented continuous curve: the only constraint we need to impose is that every variable must be a function of the individual's subjective coordinate (from now on, I will call this the "functional condition"). Let us now suppose all agents to be simultaneously present at 0, the origin of the n-dimensional Cartesian space, and let us consider the plane generated by the objective coordinate, y, and my own subjective coordinate x_1. In the future, nobody but me will have access to this space: any other agent could only reach it by reversing his temporal path and bringing back his subjective coordinate to zero, which would obviously violate the functional condition.

As a consequence, a change from event E to event E'' along the x_1 coordinate, as shown in picture 6, will be exempt from any external interference in an n-dimensional framework: the only condition is that I take care of travelling backwards with respect to everybody else's subjective direction, as well as to the objective y coordinate. Conversely, if I decide to keep following other agents' chronologies and simply move up and down along historical time, I will perform a *social* or *public* time travel since anybody else will be capable of interacting with the events I have revisited (in the picture, E').

Besides complete privacy and complete publicity, the system allows for *partially shared* time travel at all levels: if I want to involve a certain set of k individuals in a temporal experience, while leaving the others out, I will keep the formers' coordinates $\left(x_{j_1}, x_{j_2}, \ldots, x_{j_{k-1}}, x_{j_k}\right)$ increasing, and force all other x_i's to remain constant. Mathematically, travelling with k partners takes place within a $(k+2)$-dimensional subspace[18] of $\mathbf{R}^n$ (including also historical time and my own subjective direction): in the boundary cases of an entirely private or public journey, we get respectively a Cartesian plane and all of $\mathbf{R}^n$.

It is possible to show that changes brought to an event E in a given subspace S always have an effect on the image of E in every space containing S, but not conversely.

Let us first recall the constraints on scattering lines in the (x,y) plane: the horizontal direction may never be reversed with respect to subjective time, otherwise indirect vicious circles would be possible in principle; vertical scattering is equally oriented towards the future. In an interactive situation, it is reasonable to expect that the effects of an objective change to an event E will not depend on which agent performed the change, and therefore we must suppose the non-reversal condition to hold simultaneously with respect to all subjects (plus the objective timeline, once again). Algebraically, this means that an event located at $(x_1, x_2, \ldots, x_{n-1}, y)$ can only affect points $P(x'_1, x'_2, \ldots, x'_{n-1}, y')$ such that $y' \geq y$, and $x'_i \geq x_i$ for any i. Geometrically, we can say that the region of influence of the event is an infinite "cube," whose only vertex coincides with the event itself; when the latter is located at 0, the edges of the cube correspond to the positive halves of the coordinate axes. Just as we did in two dimensions, we can distinguish a dynamic and a static scattering, respectively along the objective axis and in any other direction.

---

[18] The area where I can move is not exactly the whole subspace, since I need to respect the constraint that all subjective coordinates be increasing, including mine: therefore, if the starting point is 0, the accessible region is the subset of $\mathbf{R}^{k+2}$ where $x_j > 0$ for $j = 1, i_1, i_2, \ldots, i_{k-1}, i_k$.

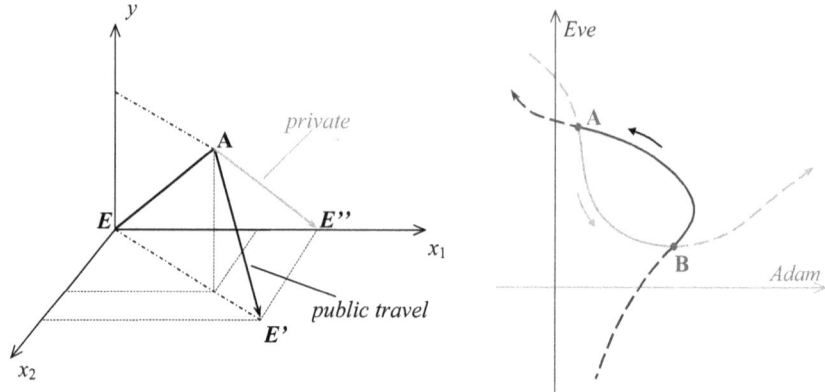

Picture 6: public and private travel

Picture 7: closed loop set up by two agents

We can now get back to the case of partially shared travel, with k subjects (indexed with $i_1, i_2, \ldots, i_k$) involved in the same temporal experience. Let us call $S(i_1, i_2, \ldots, i_k)$ the subset of R^n which they have exclusive access to: as we saw before, $S(i_1, i_2, \ldots, i_k)$ corresponds exactly to the set of points where $x_{i_1} = x_{i_2} = \ldots = x_{i_k} = 0$, and $x_j \geq 0$ for any other j. Considering all we said about scattering lines, we can immediately deduce that changes are never transmitted towards subspaces contained in S, since these would have $x_j = 0$ for some of the j's above, thus violating the functional condition.

3.2. Indirect circles

The model I have just described enables an arbitrary number of agents to perform time travel with any degree of privacy: apparently, all we need to assume is the usual non-reversal condition on subjective curves, plus the positive orientation of scattering lines in any direction.

But is this model fully consistent? We have seen that no agent alone can close his own path by making contact with his subjective past; neither can this be done through an indirect transmission of data from the future, because in that case even stricter constraints would apply—presumably, that *all* coordinates increase, as with dynamic scattering lines. It remains to be seen whether a vicious circle could be set up through the combined movement of two or more agents.

For the sake of graphic representation, I shall first consider a case with just two active subjects—let's call them Adam and Eve. Furthermore, I will initially neglect the objective coordinate and consider instead the Cartesian plane with the two subjective directions. It is easy to see that circular paths can be described by intersecting subjective curves in two distinct points, without either character violating the functional condition. This is possible because the area where Adam can move includes a set of vectors (i.e., the whole second quadrant) whose opposites belong to Eve's own access region; consequently, if he construes his curve so that the tangent vector lies constantly within the second quadrant, she will always be able to close the path in the opposite direction.

The simplest constraint which can prevent this from happening is that the tangent vector to either subject's path at any point must be contained within one and the same half-plane. This entails that whatever direction Adam takes during his course, its opposite will always lie outside the accessible half-plane, and therefore it will not be a legitimate direction for Eve at any point. Although the borderline for the half-plane is not uniquely determined, symmetry requirements suggest that we choose the secondary bisector; as a result, either agent will still be able to recede into their counterpart's past, as long as the tangent direction does not exceed a 45° angle—to make it more intuitive, Adam can travel backwards into Eve's past provided he does not move "faster" than she does in the other direction.

In order to obtain the region the first subject can reach from 0, this "diagonal" constraint must be combined with the usual functional condition (which we might call "orthogonal"): as shown in picture 9, the intersection of the two half-planes is a 135° angle having one edge on the secondary bisector, and the other on the positive direction of Eve's x_2 axis.

We can now easily revert to the complete model: since the y coordinate can take arbitrary values on both subjective curves, all we have to do is project the two-dimensional representation along the direction of that coordinate. The angle will then become a dihedron of the same spread, whose edge coincides with the y axis.

It is also possible to obtain the diagonal constraint through a more formal reasoning. First of all, let us suppose that the sum of coordinates, (x_1+x_2), be an increasing function along either subjective curve: clearly, this is a sufficient condition for non-circularity, since a closed loop (including those generated by two curves) must have the same pair of coordinates at the endpoints—and hence, *a fortiori*, the same sum.

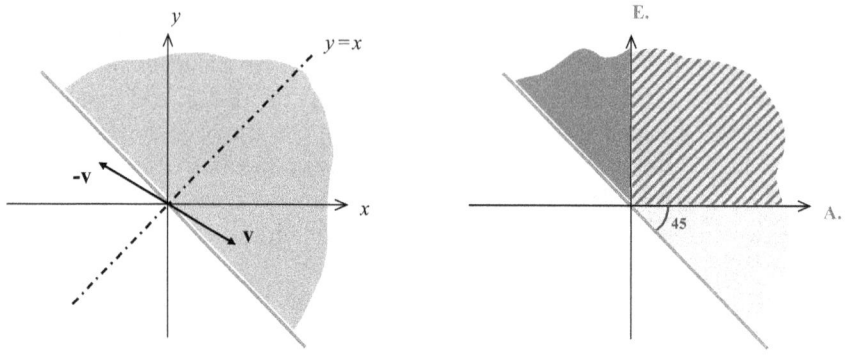

Picture 8: diagonal constraint

Picture 9: regions of access, including the diagonal constraint

Let us now look for a geometric translation of this requirement. The shortest way to do this is by considering the gradient vector of the function $f(x_1, x_2) = x_1+x_2$, i.e. the vector (1,1): it is known from multi-variable calculus that this is exactly the direction where the function has the maximum increase. Consequently, the directions with positive increase will correspond to those vectors whose inner product with (1,1) is greater than zero; since the secondary bisector is orthogonal to (1,1), the requirement means that the tangent vector at any point, for any subject, must lie on the same half-plane as the gradient, with respect to the bisector.

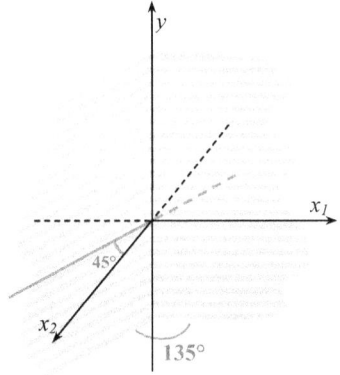 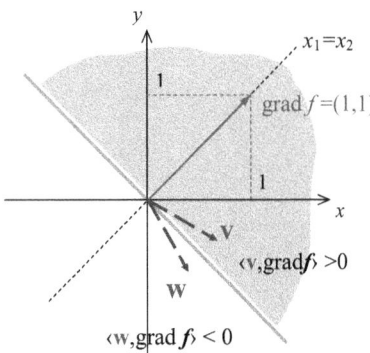

Picture 10: a complete representation of the two-subject case. The vertical coordinate is objective time; each subject can move within a 135° dihedral angle.

Picture 11: the geometric meaning of the diagonal constraint. The coloured area is the set of points where the inner product, ⟨v, grad f⟩, is positive.

3.3. Consistency constraints in higher dimensions

Though very technical in its application, the idea of the addition function is particularly useful when it comes to generalizing to any number of variables. Again, I will first consider the subspace of subjective coordinates, $(x_1, x_2, \ldots, x_m)$: since m is arbitrarily high (or even infinite), vicious circles can be generated in principle with the contribution of any number of subjects, each covering a section of a closed oriented path—or perhaps several non-consecutive sections.

In order to avoid this unusual "relay" time travel, it is sufficient once again to assume that the sum $(x_1+x_2+ \ldots+x_m)$ be increasing along every subjective curve. In the general case, the gradient vector will be $(1,1,\ldots,1)$, whose direction we may still call the "main diagonal" by analogy with the 2-D case; since the inner product has to be positive, this diagonal and the tangent vector at an arbitrary point will have to lie on the same side of the hyperplane defined by $(x_1+x_2+ \ldots+x_m) = 0$. Picture 12 exemplifies the whole situation when $m=3$.

The diagonal constraint is common to all members of the system, which is perfectly consistent with the fact that we describe it through a symmetric vector such as $(1,1,\ldots,1)$. As usual, we need to combine it with the relevant orthogonal condition in order to find the accessible region for

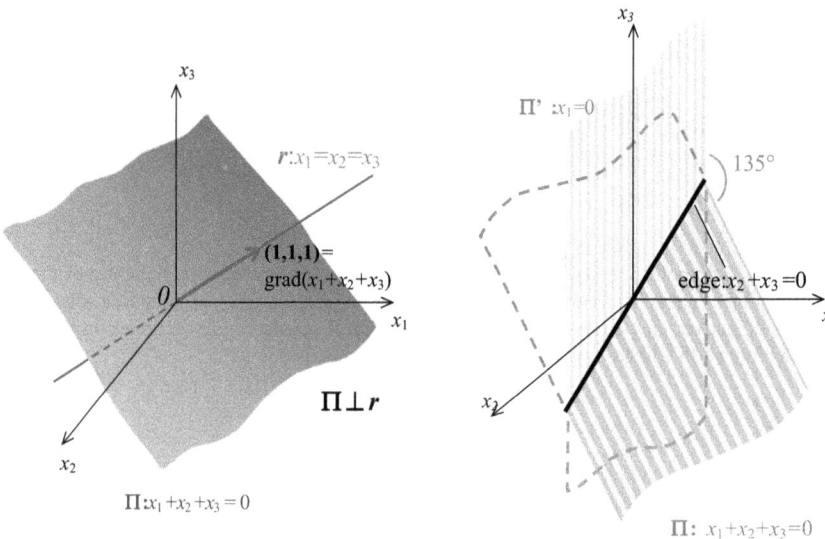

Picture 12: the diagonal constraint in the three-subject case: departing from 0, each agent's path must point outside the plane π, i.e. towards the observer.

Picture 13: the dihedron in the three-subject case

a given subject: this will turn out to be a dihedral angle, whose boundary consists of two hyperplanes intersecting in an $(m-2)$-dimensional subspace (more precisely, I will only consider half of each hyperplane). For example, supposing I take the role of the first subject in the model, the points I can reach from 0 are all those where

$$\begin{cases} x_1 > 0 \\ x_1 + x_2 + \ldots + x_m > 0 \end{cases}$$

When I alternatively turn one of the inequalities into an equation, the solutions will be the two components of the boundary region; the system with both equations corresponds to the ($m-2$)-dimensional edge of the dihedron.

Switching to the complete model is just as immediate as it was in two dimensions: as objective time travel is free, the y coordinate cannot appear in the boundary conditions; consequently, the same inequalities will hold. The geometric picture will also remain unchanged, except that an extra dimension must be added to every subspace: the result is again a dihedral angle whose boundary components are now m-dimensional; since we are moving in ($m+1$) dimensions, these will still be hyperplanes.

As for the width of the angle, we can compute it by intersecting the bordering region with Π, the unique plane which is perpendicular to both $x_1=0$ and $x_1+x_2+\ldots+x_m=0$: this is the plane generated by the ($m+1$)-dimensional vectors $(1,0,0,\ldots,0)$ and $(1,1,1,\ldots,0)$, where the first m components are equal to 1. It is convenient to take orthogonal generators, such as $v_1=(1,0,0,\ldots,0)$ and $w_1=(0,1,1,1,\ldots,0)$: if we call v_i the unit vector along the i-th subjective direction, the two coordinate sequences will correspond to v_1 and $w_1 = v_2 + \ldots + v_m$, respectively.

It is worth remarking that the y coordinate is not a component of either generator of Π: this means that, for any choice of m, the width of the angle will not depend on whether we add the objective axis or not (we already knew this for $m=2$).

Let us now consider the projection of the accessible region on Π. As we know, the first boundary component has $x_1=0$ and $x_1+x_2+\ldots+x_m > 0$: its intersection with Π will be the top half of the vertical axis. On the second component, the sum has to vanish, and therefore we may write $x_1 = -(x_2+\ldots+x_m)$, i.e. we must take the vector sum of v_1 with the opposite of w_1 (the right half-line, since $x_1>0$).

As a result, the angle turns out to be greater than 135°. This happens because w_1 is not a unit vector: its length being $\sqrt{m-1}$, the spread of the dihedron will be exactly $90°+\text{arctg}\sqrt{m-1}$; the value tends to a straight angle when the number of subjects increases to infinity.

4. Open questions

With this result, the geometric model is apparently complete. Not only have we ensured the availability of a private past in frameworks with an arbitrary number of time travellers, but we have brought up the existence of a whole range of options between private and public travel, providing any given set of agents with the possibility of intervening collectively on a certain past event. Besides, the diagonal constraint allows escaping both direct and indirect circular loops, which would make the whole system logically untenable.

Despite this, the investigation is far from being over, for a further generalization is in order, concerning the mutual position of subjective curves. When I discussed the possibility of private time travel, I required that all subjects be initially located at the origin. This is a special configuration that cannot be presupposed in general, since the agents' temporal movements are by hypothesis unrestrained; once the requirement is lifted, portions of time that were previously private—perhaps all such portions—mightbecome accessible to everyone. If this is the case, further mathematical tools will have to be devised in order to restore the privacy condition, some of which might involve a conception of agents as higher-dimensional objects such as smooth surfaces or varieties, rather than temporal curves.

The generalization will have a bearing on some of the interpretive problems I have previously hinted at, as well as bring about new ones. Thus, for instance, the assumption of a common departure point has an obvious connection with that of an *absolute present*, i.e. the bisector of the first quadrant, where objective time and the different subjective chronologies proceed at the same pace: once we revoke the special status of the origin, it seems difficult to preserve it for the bisector. Moreover, both issues are relevant to the relation between an agent and his counterparts: provided we still view individuals as curves, the existence of singular points or directions in time could offer an argument for choosing a specific curve out of a sheaf and bestowing an ontological privilege upon it (although the privilege may as well be sustained on different grounds).

Another problem that may descend from relinquishing the idea of an absolute present concerns the possibility for agents to communicate, that is, to intentionally cross each other's path: should direct contact turn out to be unwarranted, an alternative—and once again, mathematics comes into the picture—is to develop a notion of temporal action at a distance, perhaps with some proximity constraints.

The final outcome of the inquiry need not be a straight yes or no to the many-dimensional topology: the story might as well be coherent but beset with limitations and auxiliary conjectures. In that case, once the formal work is over, there would remain the philosophical problem of weighing these conditions against the intuitions and intellectual needs that are behind the hypothesis of time travel, and decide if the price is fair and worth to be paid.

REFERENCES

Ballard, J. (1956). Escapement. *New Worlds*, ed. J. Carnell, 54. London: Nova Publications, 127-139.

Bars, I., Deliduman, C., and Andreev, O. (1998). Gauged Duality, Conformal Symmetry and Space-time with Two Times. *The Physical Review*, D58, 066004.

Broad, C. D. (1937). The Philosophical Implications of Foreknowledge. *Aristotelian Society Supplement*, 16, 177-209.

Chen, X. (1999). A New Interpretation of Quantum Theory—Time as a Hidden Variable. URL: <http://arxiv.org/abs/quant-ph/9902037>.

Chen, X., (2005). Three Dimensional Time Theory: to Unify the Principles of Basic Quantum Physics and Relativity. URL: <http://arxiv.org/abs/quant-ph/0510010>.

Goddu, G. C. (2003). Time Travel and Changing the Past: (or How to Kill Yourself and Live to Tell the Tale). *Ratio*, 16, 16-32.

Grey, W. (1999). Trouble for Time Travel. *Philosophy*, 74, 55-70.

Grimwood, K. (1987). *Replay*. Westminster: Arbor House.

Hanley, R. (2004). No End in Sight: Causal Loops in Philosophy, Science and Fiction. *Synthese*, 141, 123-152.

Hawking, S.W. (1992). Chronology protection conjecture. *The Physical Review*, D 46(2), 603-11.

Horwich, P. (1987). *Asymmetries in Time*. Cambridge (US): MIT Press.

Jameson, M. (1941). Doubled and Redoubled. *Unknown Fantasy Fiction*, ed. J. W. Campbell, IV (5), 87-99.

Levin, M. (1980). Swords' Points. *Analysis*, 40, 69-70.

Lewis, D. (1976). The Paradoxes of Time Travel. *American Philosophical Quarterly*, 13, 145-152.

Meiland, J. W. (1974). A Two-dimensional Passage Model of Time for Time Travel. *Philosophical Studies*, 26, 153-173.

Moore, G.E. (1912).*Ethics*. London: Williams and Norgate.

Oaklander, L. N. (1983). McTaggart, Schlesinger, and the Two-dimensional Time Hypothesis. *Philosophical Quarterly*, 33, 391-397.

Smart, J.J. (1949). The River of Time. *Mind*,58 (232), 483-494.

Vallée, R. (1991). Perception, Memorisation, and Multidimensional Time. *Kybernetes*, 20(6), 15-28.

Weinstein, S. (2008). Many Times. URL:<http://arxiv.org/abs/0812.3869>.

GÖDELIAN TIME TRAVEL AND WEYL'S PRINCIPLE*

VINCENZO FANO AND GIOVANNI MACCHIA
Università di Urbino, University of Western Ontario
vincenzo.fano@uniurb.it, gmacchi@uwo.ca

ABSTRACT. We discuss an aspect that the philosophical debate concerning time travels (TTs) seems to have overlooked. This debate usually takes into account two kinds of TTs: *Wellsian* (WTTs) and *Gödelian* (GTTs). In the former, travelers can roughly be imagined suddenly disappear at a certain instant and then somehow reappear at another instant. In the latter, travelers simply start at a given point in spacetime, take a conventional trip, and finally return to that very point. In WTTs, the travelers worldlines have anomalous structures that are not easily explainable from a physical viewpoint, leading, furthermore, to indigestible causal relations in which effects precede causes. For these reasons WTTs are often not considered deserving serious scientific attention. GTTs, on the contrary, result from the physics itself, as solutions of the Einstein field equations. The particularity of these solutions resides in the topological structures of their spacetimes, so curved as to mould the so-called *closed timelike curves* (CTCs). They are continuous worldlines devoid of *global* temporal ordering inasmuch as each point both precedes and follows all the others (although *locally*, on a neighborhood of each point, events along CTCs are ordinarily ordered from past to future). Accordingly, WTTs, occurring in a globally ordered temporal background, can be seen, in a sense, as the "true" TTs insofar as they are travels *through* time, "movements" over time. GTTs instead, not occurring in such a background, are not strictly TTs but are, as it were, travels "of" (space)time: their CTCs represent the very spacetime journey itself (not just the spatial path on which traveling over time). However, there are "*local* cases," containing CTCs in a limited region alone, in which GTTs can recover an important Wellsian character. One of these local cases is the Kerr-Newman spacetime, describing particular black holes whose existence is more and more believed to be possible. In our universe, the existence of a *cosmic time* is deemed possible too. Indeed, an important foundational principle of modern cosmology, *Weyl's Principle*, allows to foliate spacetime in a sequence of "space slices." Thus it makes sense to speak of a temporal evolution of the universe as a whole because a global temporal order with respect to the relations "later than" or "earlier than" exists. Consequently, universes respecting Weyl's Principle make possible, in principle, to compare local time orders of CTCs with temporal orderings of neighbouring and distant worldlines, all existing in a sort of "absolute" temporal standard of reference, thereby recovering for the mostly physical GTTs, at least partially, a more "natural" sense of journey *through* time.

---

* We thank Silvio Bergia for a first sketch of the idea on which this paper is based.

1. Introduction

Time travel (TT) stories have always been a staple of fantasy literature. There is no real agreement among scholars as to which written work is the earliest example of a TT story, the origin of the notion itself being somewhat ambiguous, but it seems that something akin to travels *forward* in time can even be recognized in ancient folk tales and myths (some centuries BC), whereas travels *backward* in time seems to be a more modern idea (around the eighteenth century).[1] Obviously, the prospect of building a machine in which one could be transported through time has for the most part belonged to the realm of fantasy. Over the last few years, however, issues concerning TTs and time machines have also received a great deal of attention both in the physical and in the philosophical context, thus becoming part of modern scientific inquiry. The literary pedigree of these matters has nevertheless not been completely expunged from these more recent studies. This is evident, for example, from the name itself given to one of the two commonly analysed varieties of TT, the so-called *Wellsian time travel* (WTT), that obviously derives from Herbert George Wells' 1895 milestone *The Time Machine*.[2] As is well known, in his tale Wells conjectured that a time machine could travel through time in the way a simple car can travel through space. These journeys, therefore, consist of "motions" in *both* temporal directions, even though the time machine remains in its initial spatial location.

It goes without saying that the cultural landscape concerning TT has completely changed since 1895. However, the fact that this tale, its author, and variegated quotations from other literary and cinematographic works sometimes peek out from physical and even more so from philosophical essays provides pleasing evidence of how TTs and time machines, though they are now difficult subjects impregnated with physical-mathematical technicalities and philosophical subtleties, also continue to be fascinating mysteries still deeply rooted in our imagination.[3] But, for various good reasons that we shall expound later, WTTs have never been considered wor-

---

[1] For an historical outline on TT (and relative references) see the entry "Time Travel" in the online encyclopedia Wikipedia.

[2] Meiland 1974 is maybe the first author to have made this association and used this label.

[3] For a delicious example of such a rich blend of literature, cinematography, comics, mathematics, physics and philosophy, all obviously flavored with conspicuous doses of human fantasy, see Nahin 1999.

thy of scientific interest. However, at least until 1949, they were thought of as the paradigm of a TT story. It was on that date, in fact, that Kurt Gödel elevated the scientific legitimacy of the discussions on TT thanks to the discovery of a family of rotating universes in which hypothetical travelers can freely voyage, under suitable conditions, between any points in spacetime, either returning to the same (in space *and* time) starting point or to a preceding one. This new kind of TT, called *Gödelian time travel* (GTT), is completely different from the Wellsian one insofar as a GTT occurs (in a theoretical sense, of course) by taking advantage of the geometrical structure (the curvature) of spacetime itself, so that its hypothetical feasibility might become stronger by being rooted directly in the mathematical structures of the physical theories. For this reason GTTs are preferred by scientists and by most philosophers.

In this way, having on the one hand a sort of "literary" TT and on the other a sort of "physical" one, naturally the dispute between their plausibility, at least from a logical-conceptual point of view, is undoubtedly in favour of the latter. However, even if GTT is rightly considered as deserving much more scientific attention—and, moreover, it is often considered as the paradigmatic and unique example of what a TT should be in order to have a sense—in our opinion such type of travel is not completely satisfactory. Indeed, a TT *tout court* needs a constitutive ingredient that intrinsically belongs to a WTT but not to a GTT. And this ingredient is a global ordered time in which the travel, still before occurring, might conceptually constitute as an effective journey *in* time, something not allowed by the spatiotemporal background of Gödelian travel.

The following pages are dedicated to the explanation of such an idea. We shall try to show why and how that ingredient may be naturally added, or rather recovered, also in some special cases, but not of secondary importance, of GTT.

The paper proceeds as follows.[4] Section 2 presents a general definition of TT. Section 3 is a survey on the two types of WTT and their conceptual weaknesses. After having briefly introduced, in section 4, the causal structure of general relativistic spacetimes, in section 5 the charac-

---

[4] A premise: we shall not enter into analyses of TT paradoxes, nor into results obtained by physical theories other than General Relativity, nor into more general issues concerning the compatibility between TT and the contemporary philosophies of time like presentism, eternalism, endurantism, and so on. Not because all these are not interesting matters, but simply because the central point of our discussion does not concern them directly.

teristics of GTT are investigated, in particular those of the travels in Gödel's universe. Both in section 6, in which the advantages and disadvantages of Wellsian and Gödelian journeys are compared, and in section 7, where their status as TTs is confronted, we discuss in which sense a GTT cannot be considered properly a TT. In section 8, two examples of *local* GTTs, namely occurring in bounded regions of spacetime, are analysed, while a third example, regarding Kerr-Newman black holes, paves the way for section 9. The latter introduces the so-called *Weyl's Principle* and it is then argued how some local GTTs—provided that in their universe such a principle holds, so that a global ordered temporal background is recovered—could be more properly regarded as actual TTs.

2. What is time travel?

Perhaps the most used definition belongs to Lewis. He wonders:

> What is time travel? Inevitably, it involves a discrepancy between time and time. Any traveler departs and then arrives at his destination; the time elapsed from departure to arrival (positive, or perhaps zero) is the duration of the journey. But if he is a time traveler, the separation in time between departure and arrival does not equal the duration of his journey. (Lewis 1976, 145)

Lewis highlights that in order to give sense to the concept of TT one needs to distinguish the *personal* (or *proper* or *subjective*) time of the traveler from the *external* time of his/her surroundings. According to Lewis, in order to have a TT, personal and external times must not march in step (as instead happens for non-time travelers): one travels in time if one traverses some temporal interval in the proper time that differs from the duration (in external time) of that interval.

Consequently, such a definition implies that TTs are always incompatible with the so-called *chronological monism*, according to which the subjective time arrow of the traveler necessarily coincides (modulo relativistic discrepancies) with the time arrow of the external world so that only one time series exists. This incompatibility is manifest: when the travel is directed into the past, the traveler clock moves in the opposite direction with respect to the external time (however, this does not imply any discrepancy in temporal lapses—contrarily to what Lewis says[5]—insofar as a

---

[5] Horwich (1987, 114) makes the same mistake.

traveler may experience a journey lasting an hour of his/her proper time but arriving one hour before with respect to the external time); when the journey goes into the future, even if the direction of both time arrows is the same, the traveler's clock must necessarily "jump ahead" with respect to external time. Thus, in both cases the duration of the journey and the external time elapsed between the traveler's departure and arrival either are not spanned by a single stretch of time or have a different arrow direction or both.

It follows then that Lewis' definition, as Smeenk and Wüthrich (2011, 580) rightly note, does not grasp a necessary condition for TTs because, as already stated, a backward TT could simply occur even if its duration *equals* the temporal separation (measured in external time) between departure and arrival. Not only, and more significantly: within the context of modern physics such a definition is not even a sufficient condition. The reason is that the meaning of the Lewisian locution "separation in time between [...]" is not at all clear in the context of Special Relativity, insofar as the relativity of simultaneity in general implies different separations between events for observers in relative motion.[6] And to the best of our knowledge Special Relativity holds.

Other authors depart significantly from Lewis' definition adopting, in a way, a simpler one, whose most important characteristic is that it is drawn from some solutions of general relativistic equations. According to this definition a TT is simply a journey happening on a closed timelike curve. Before introducing this kind of curve and the whole meaning involved in this concept, and, more in general, the perspective leading us to attribute the status of "physical" TT only to GTT, it is first necessary to present in more detail the WTT, and not by chance this kind of TT fits better Lewis' aprioristic definition.

3. Wellsian time travel

The "dynamics" of a classical WTT, which is usually considered to take place in Newtonian or Minkowskian spacetimes, is simple: the usual traveler with his/her time machine, persisting in a normal way through time, is imagined to suddenly disappear *ad nihilum* at a certain instant and then somehow reappear *ex nihilo* at another instant (not necessarily in another

---

[6] Two proposals, both failing, to assign an objective meaning to that locution are analyzed by Smeenk and Wüthrich (2011, 580).

place). Our traveler, therefore, moves backwards (or forwards)[7] in a kind of "spatialized time."[8] In this kind of TT it is evident that worldlines of time travelers and causal relations among the events on these worldlines are particularly, as it were, unusual (worldlines, for instance, break off abruptly, but in the next sections we explain this better), whereas spatio-temporal and causal structures are "standard," in the sense that, as is well known, in these spacetimes all light cones are aligned so that a unique direction may be assigned to time, and then nothing strange, intrinsic to spacetime itself, happens regarding causal and temporal relations.

WTTs have two variants both involving a peculiar worldline structure: in one case the traveler jumps *discontinuously* from a later time to an earlier one (as measured in the external time), in the other he/she gradually and *continuously* make his/her way back.

3.1. Discontinuous Wellsian time travel

Fig. 1 (readapted from Earman 1995, 162) illustrates the first case of a journey through a discontinuous worldline. At the top, the traveler T_1, in a time machine, cruises along until dialing, at e_1, the target date into the counter to "minus 300 years" and pressing the button, causing the traveler and the time machine to disappear, while time is "rewinding" until the target date is reached. At e_2, i.e. three hundred years prior to e_1[9], a person T_2, exactly resembling T_1, both in terms of physical appearance and mental states, instantly pops into existence. The worldline of the time machine/traveler, stopping at the instant e_1 and restarting at the earlier external time e_2, is thus discontinuous. Such a restarting, from the standpoint of the traveler's proper time, can be instantaneous or not, and in any case, between e_1 and e_2, the time machine and its occupant pass into non-existence

---

[7] Hereafter we ignore, for the sake of simplicity, the future-directed WTT. On the other hand, in general the philosophical interest for this kind of trip is, in a sense, comparatively lesser since Special Relativity already teaches us that it may be accomplished by moving at a very high speed and therefore slowing down aging (the famous *twin effect*).

[8] We can roughly regard the notion of "spatialized time" as "an extended something along which we can move" (Smart 1955, 241). Here Bergson's concept is not at stake.

[9] We are measuring in Newtonian time, or in the inertial time of the frame in which T_1 is at rest.

as they do not occupy a spatiotemporal place (from a standpoint external to the traveler).[10]

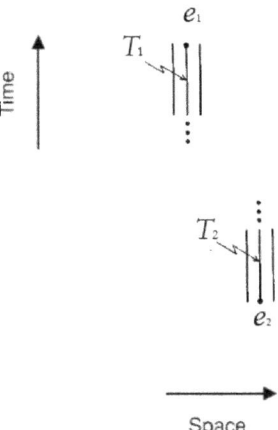

Fig. 1. Discontinuous Wellsian time travel.

In order to be regarded as a true journey in time, this hypothetical sequence of events urges another fundamental assumption, so far merely hinted at: it is necessary that, although T_1 is discontinuous with T_2, the person who arrives at e_2 is the *same* person who departed from e_1, namely that T_2 "is in some appropriate sense a continuation of" T_1, as Earman (1995, 162) points out. In other words, *genidentity* (i.e. the identity through time) should be preserved and a causal continuity is also required according to which events on T_1 cause the events on T_2 (the disappearance itself of T_1 at e_1 should be considered as the cause of the appearance of T_2 at e_2).[11] Even

---

[10] This is the classical criterion for the existence of a material object. Obviously, one may also refuse it.

[11] How to determine whether different stages at different times belong to the history of the same object is a very difficult problem. The concept of the individual that remains identical during the passage of time has been denoted by the German philosopher Kurt Lewin in 1922 by the term *Genidentität* (genidentity), or identity through change. Many philosophers have provided different criteria of genidentity. Here we are adopting a slightly modified version of the most influential one, proposed by Reichenbach and Carnap. Genidentity is an equivalence relation, established by a continuity between world points, or moments of particles, holding in *either* temporal direction. In the words of Reichenbach: "The points of a timelike worldline [...] are referred to as states of the *same* object" (1957, 270), and Carnap: "Two world points of the same

though in the order of the time traveler causation runs normally from earlier to later stages of his/her proper time, the point is that such a personal order disagrees with the external time order. And the unpleasant consequence is that in this latter order there will be reversals of the direction of the causation action in which the effects precede the causes, that is, there will be causal relations backward oriented from future to past. This is the so-called *backward causation*, a phenomenon not at all physically attractive, or, even worse, completely unphysical for most authors and for our scientific common sense.

However, it is really difficult, if not impossible, to eliminate backward causation if we want to read such a sequence of events as a TT. We argue for this in the following section.

3.1.1. *Discontinuous Wellsian time travel* denied

Consider another interpretation of these same facts. Imagine a different scenario in which a sort of conservation principle of mass-energy is respected in the following way: at e_1 the dematerialization of the system traveler/time machine is compensated by an instantaneous materialization, still at e_1, of an equivalent amount of energy, while at e_2 a non-material form of energy is converted into an equivalent amount of matter (clearly, we are referring to the matter constituting the machine and its occupant). Even if, physically speaking, such a situation is very strange, the interesting point is that the causal link going from e_1 to e_2 is really weakened, or even completely nullified, insofar as nothing causally relevant seems to go *backwards* from e_1 to e_2: the causal resultants of e_1 can be traced *forwards* in external time (indeed, after the disappearance of T_1 nothing goes back to e_2), and the causal antecedents of e_2 can be traced *backwards* in external time (nothing arrives from e_1, i.e. from the future), everything, in short, according to the dictates of a normal causation. Accordingly, backward causation is nullified, but now genidentity seems lost too (no "information" is transferred between e_1 and e_2) and the same happens for the TT itself. Note that it is true that this interpretation appears more physical because at least a conservation principle of energy is respected and backward causation is

---

world line, we call *genidentical*; likewise, two states of the same thing" (Carnap 1967, 199). Besides, their criterion, in accordance with modern physics, closely relates the concept of genidentity to the concept of causality as Reichenbach explicitly affirms: "Different states can be genidentical only if they are causally related" (1957, 271).

eliminated, but phenomena like dematerializations and transformations of mass in energy (and vice versa) like the ones seen here, are quite obscure for present day physics, not to mention the "resemblance," at this point truly incredible and unjustified (pure coincidence?), of T_2 to T_1.

So, if we get rid of backward causation we risk doing away with TT itself insofar as an interpretation that favours them is no longer necessary; on the contrary, if we want to preserve TT status, and thence the genidentity included in it, we have to digest backward causation.

3.2. Continuous *Wellsian time travel*

The most simple representation of this kind of TT is given by a N-shaped worldline that "bends backward" on itself as in fig. 2(a). In this representation, our traveler T_1 in a time machine cruises along its worldline (the arrows indicate increasing biological time of the traveler). At e_1 the time machine turns backward in time, whereas at e_2 it turns forward in time again.

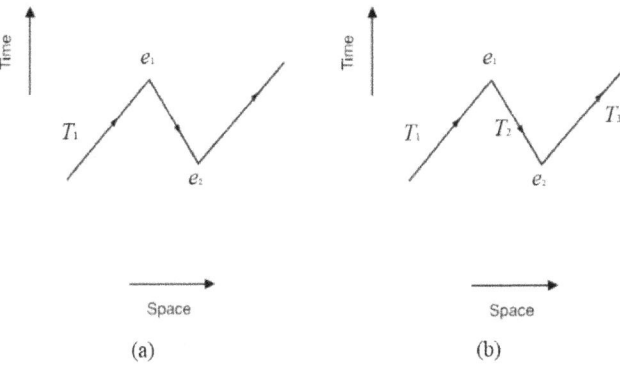

Fig. 2. Continuous Wellsian time travel (a).
In (b) no Wellsian time travel takes place.

The worldline of the time machine is evidently continuous (no disappearances are involved) even if some events belonging to its "second part" (from e_1 to e_2) occur, with respect to some events of the "first part" (before e_1), at an *earlier* external time (whereas, for the traveler's proper time, *all* the events of the second part are subsequent to those of the first part). In other words, from the point of view of an observer living in the external time all the processes in the time machine are going backward in time between e_1 and e_2, whereas from the traveler's viewpoint the external world is going backward in time and nothing strange happens inside the machine.

This interpretation of the facts involving a TT seems to require a causal significance for the arrows on the worldline segments, in other words—as in the first interpretation of the discontinuous WTTs—the preservation of genidentity implies the backward causation (later events, before e_1, cause earlier events, after e_1, both with respect to external time). As regards the ontological status of the traveler (and the time machine), it again results in a crazy situation, i.e. a sort of tripartition insofar as, at any instant between e_1 and e_2, three travelers ("images," it seems, of the *same* person but of different ages from the traveler's standpoint) exist at the same time (one going forwards, the other (older) traveling backwards, and the last (still older) persisting again forwards). In short, our traveler would have the gift of a particular ubiquity, being present in several places at the same time, moreover endowed with different ages. Furthermore, were the time machine to be spatially at rest, even a sort of spatial coincidence would be implied, because those three travelers would share the same spatial location.

3.2.1. Continuous Wellsian time travel denied

Also in this case it is possible to view the sequence of events in a way in which backward causation is denied, but again all this at the expense of TT itself. The representation of this emended scenario starts from the event e_2 (see fig. 2(b)), where it is possible to imagine that a pair of middle-aged twins, T_2 and T_3, is spontaneously created. The second twin is the "normal" one, aging in the ordinary biological way, while the first, on the contrary, gets progressively younger. But there is another person, T_1 (temporal mirror image of T_2), that in the meantime is aging normally until meeting T_2 at e_1 where they annihilate one another. This physically very strange interpretation, even if it is still adapted to the continuity of the entire worldline, evidently breaks up the genidentity because we no longer have only one person but three *different* ones.[12] In this framework TT itself disappears.

Summing up, both sequences (continuous and discontinuous) of events, when considered as TTs, somehow require a preservation of genidentity and a continuation of the causal chain regarding the traveler, so according-

---

[12] See a more detailed description of this scenario in Putnam (1962, 665-668).

ly both types of WTTs are bound up in an inextricable way with backward causation. But, if we are not available to swallow such a physically indigestible reversed form of causation, we have to renounce TTs themselves. In any case, whatever our choice, it will end by implying some inexplicable physical phenomena, although not contradictory.[13]

It is hopefully clear why WTT is not considered worthy of serious scientific attention.

4. The physical foundations of Gödelian time travel

In order to introduce GTT and more general topics regarding the possibility of TT in general relativistic spacetimes it is propaedeutic to offer at least a brief survey of their basic causal structure describing the causal relationships among spatiotemporal points (roughly, which events can influence which other events). As is well known, the concepts of chronology and causality in General Relativity present many more problems that in Newtonian and in special relativistic physics. In these last two cases, in fact, the nature of their spacetimes clearly separates past from future. But in General Relativity, Einstein field equations are *local* equations that relate in each point some aspects of the spacetime curvature to the general distribution of matter-energy. Therefore, it can happen that some spacetimes, being particularly curved by matter, generate on a global scale the so-called *closed timelike curves* (CTCs), i.e. continuous and differentiable worldlines devoid of a *global* temporal ordering inasmuch as each point both precedes and follows all the others. This means that an object moving along such a curve towards its local future would travel back into its own past to the very moment and place at which it began its journey. Note that *locally*, on a neighborhood of each point, a CTC does not differ from the *open* worldline of an ordinary material particle (in a flat spacetime), the events of whose story are ordered along this worldline from past to future. The peculiarity of General Relativity is that it does not impose *global* constraints on its spacetimes. In cosmology, for example, the spatial topology of the universe has to be discovered experimentally. And the global temporal topology, in which we are particularly interested here, is not constrained by Einstein field equations either. Therefore, it does not matter if the spacetimes

---

[13] We remind them all together: materializations and dematerializations, matter-energy transformations, annihilations, people getting younger, spatial coincidences, simultaneous coexistences of the same person of different ages.

of General Relativity, being locally Minkowskian, locally respect all of the causality constraints of Special Relativity, because what locally may seem physically reasonable, may globally turn into "pathological" topologies, which give rise to serious worries regarding causality and chronology.

A very rich hierarchy of causality conditions can be imposed on relativistic spacetimes, but we shall mention only the main instances, using pedantic technical language as little as possible.

The first most basic concept is that of *temporal orientability*. A spacetime M, g_{ab}[14] is temporally orientable iff there is at each point a well-defined local sense of past and future, in other words iff all light cones admit a *continuous* division into two classes: the "past" and the "future" lobes.[15]

Before introducing the second causality condition, i.e. chronology, we need a definition: for $p, q \in M$, we say that p *chronologically precedes* q (in symbols $p \ll q$) iff there is a smooth future-directed *timelike* curve from p to q. Now we may say that a spacetime (time oriented) M, g_{ab} exhibits *chronology* (or *time order*) iff there is no $p \in M$ such that $p \ll p$, that is the relation of chronological precedence is irreflexive. In other words, for such a spacetime, CTCs are forbidden. A spacetime in which $\forall p \in M$, $p \ll p$ is termed *chronologically vicious*.

Another introductory notion: for $p, q \in M$, we say that p *causally precedes* q (in symbols $p < q$) iff there is a smooth future-directed *non-spacelike* curve (also called *causal curve*) from p to q. That is, the relation of causal precedence, at variance with that of chronological precedence, could be instantiated along a lightlike curve as well. The third causality condition, the *simple causality* (or *causal order*), is similar to the preceding one but more restrictive: it requires that there are no *closed* (future-directed) *causal curves* (CCCs), namely both CTCs and closed *lightlike* curves are forbidden. More formally: M, g_{ab} (time oriented) exhibits simple causality iff there is no $p \in M$ such that $p < p$, that is the relation of causal precedence is irreflexive as well. Thus chronology looks at causal influences transmitted along timelike intervals (with proper time finite), whereas simple causality also takes into account influences transmitted along lightlike intervals (with proper time null).

---

[14] Remember that M is a four-dimensional manifold representing the totality of all point-event locations, and g_{ab} is a semi-Riemannian metric of Lorentz signature on M representing the metric structure of spacetime.

[15] Which lobe is the "future" or the "past" is the problem of the direction of time that we do not cover here.

The fourth condition, *strong causality*,[16] is obtained iff M, g_{ab} (time oriented) possesses no *almost* CCCs, where an almost CCC is a causal curve which does not intersect itself (as instead a CCC does) but enters at least twice in the same infinitesimal neighborhood.

The fifth causality condition is the *stable causality*: M, g_{ab} (time oriented) is causally stable iff no slight perturbation in its (mass-energy distribution, and hence in the) metric, produces CCCs. An important theorem by Hawking (1968) states: in a spacetime a *global time* exists iff the spacetime is causally stable. The notion of global time is very important in order to understand next sections as well. A *global time function* is a differentiable map $t: M \to \mathbb{R}$ such that if $p \ll q$ then $t(p) < t(q)$. That is, such a function t "transforms" the chronological precedence relation between two events in a causal precedence relation, preserving the order. Now it is possible to define the global time t by functions monotonically increasing along *every* future-directed causal curve. Its level surfaces "$t = $ const." partition spacetime in *global time slices*, i.e. spacelike hypersurfaces, without edges, of simultaneous events. For instance, Newtonian and Minkowskian spacetimes are partitioned by such slices. In cosmology, global time is also called *cosmic time*, because it may be thought of as measuring the time that would cover the entire history of the universe considered as a whole.

If stable causality gives the best available guarantee that there will be no causality violations, the best guarantee that causality, in the sense of determinism, has a fighting chance on the global scale is instead guaranteed by a stronger condition, the sixth and last: *global hyperbolicity*. M, g_{ab} is globally hyperbolic iff it possesses a *Cauchy surface*, i.e. a three-dimensional spacelike hypersurface $S \subset M$, such that S is intersected exactly once by every timelike curve without endpoint. In such a spacetime M is topologically equivalent to $S \times \mathbb{R}$, where $S \times \{t\}$ is a Cauchy surface $\forall t \in \mathbb{R}$. Thus spacetime is partitioned by Cauchy surfaces (each level surfaces $t = $ const. of the global time function is Cauchy) and obviously it does not contain CCCs.[17]

---

[16] Actually, for the sake of brevity, we are skipping a weaker condition called *future and past distinguishing* (see Earman 1995, 165).

[17] We have said that global hyperbolicity deals with determinism because to say that a spacetime is globally hyperbolic is the technical way of saying that it does not contain pathologies (for instance, singularities, CTCs, etc.) that prevent the implementation of global Laplacian determinism. Indeed, in the so-called *Cauchy problem* (an extension of initial value problem) consisting of finding the solution to a differential equation

Summing up, global hyperbolicity implies stable causality that implies strong causality, that in turn implies simple causality, that finally implies chronology. As will become still more clear in the following sections, there is a subtle interplay between these causality conditions and the topology of spacetime as well as the emergence of a cosmic time in our universe. For the moment we have seen: 1) that local time (time arrow apart) always exists in spacetime; 2) that chronology and the simple causality condition forbids time to be closed, whereas the strong condition improves safety margins for the non-existence of such temporal loops; 3) that stable causality allows a stability of space and time measurements, and turns out to be equivalent to the existence of a cosmic time; 4) and finally that global hyperbolicity provides spacetimes with deterministic properties almost as strong as those of Newtonian mechanics.

5. *Gödelian time travel*

We have seen that WTTs can be summarised by roughly saying that they are carried out by odd worldlines in normal spacetimes. The characteristics of the GTTs are exactly the contrary: spacetimes have a peculiar causal structure that constrains the closure of the worldlines (in other respects completely normal) so that the travelers following those worldlines have to come back to the same spacetime location from whence they left. The particularity of these spacetimes resides in their topological structures, which are so curved as to mould those CTCs seen in the preceding section. Needless to say, for such a reason Newtonian and Minkowskian spacetimes are excluded as candidates for GTTs.

What is a GTT then? It is

> nothing more, and nothing less, that the act of starting at a particular point in spacetime, taking an otherwise conventional trip, and somehow returning to (or close to) that very point. No discontinuous motion is involved. Neither is superluminal velocity. In

---

which satisfies some given conditions, once initial data are specified on the Cauchy surface, an important theorem on existence and uniqueness of solutions allows us to "project" these data into the future and into the past determining *throughout* the whole spacetime a *unique* solution, namely a unique spatial location at a given instant. Instead, such an attempt is liable to fail in some region of M if S is not a Cauchy surface because that theorem would no longer be satisfied.

geometric terms, my time traveler is simply one whose worldline is closed (or almost closed). (Malament 1984, 91)

In such a way, in GTTs ambiguous forms of causality are no longer at stake, because time comes back on itself independently of the causal connections. A Gödelian traveler, indeed, is *always* oriented towards his/her local future and following this direction eventually returns to the past from whence the trip started, or even before. Thus no backward causation is involved, at least not in the form arising in Wellsian stories of TT, because CTCs are always curves directed towards *their* future. Consequently, these curves implement only a standard form of causation, namely aligned with the local direction of their proper time, thus pointing to the future. On the other hand, as Earman (1995, 164) specifies, the standard relativistic assumption—i.e., all causal influences, in the form of energy-momentum transfers, propagate forward in time with a speed less than or equal to that of light—is consistent also with CTCs, and, even if one could conjecture other kinds of causal influences not involving energy-momentum transfer, the point is that GTTs do not have any need to claim a backward causation of any type because, at least locally, the form of causation in which they are engaged is always forward.

Furthermore, it is important to remark that contrary to what happens in WTTs, on CTCs no uncanny dematerializations or materializations happen, so that the most important feature of a CTC is its continuity and therefore its natural conservation of genidentity.

5.1 Time travel in Gödel's universe

The most famous example of GTT occurs in Gödel's solution to the Einstein field equations. This solution describes the gravitational field of a uniform distribution of rotating dust matter, where, loosely speaking, the gravitational attraction of matter and the added attractive force of a negative cosmological constant Λ are compensated by the centrifugal force of rotation. Gödel's infinite universe is unusual in several respects, the principal one being that the rotation tilts the light cones, creating CTCs (see fig. 3, from Richmond 2006, 9).

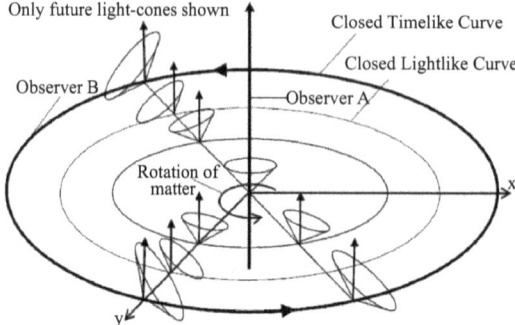

Fig. 3. Gödel's universe (one dimension suppressed).

In short, its general properties are the following: metric is stationary, not expanding; the entire material content is everywhere in a state of uniform, rigid rotation; the manifold is topologically simple: $M = \mathbb{R}^4$, thus spacetime is temporally orientable but chronologically vicious; there is a future-directed CTC through *every* spacetime point, but these curves are *not* geodesics so that TTs in a state of free fall are not possible: an accelerating "rocket" is needed.[18] The last, but most important characteristic, at least for our upcoming considerations, is that in this universe there does not exist a single global time slice, namely there is no such thing as Gödel's universe at a certain (global) time, so time does not have a unique direction.

One interesting consequence of this fact is underlined by Richmond (2006, 9). He notes that Gödel's universe implies that different observers have time axes orthogonal to each other, like, for instance, the observers A and B in the fig. 3. The cases of A and B are completely symmetrical: A records the whole life of B as passing in an instant of A's proper time, and the same happens for B when looking at A's life.[19] Richmond significantly concludes affirming: "Such a symmetry seemingly destroys any chance of claiming that there is any (at least physical) basis for claiming that one of the two observers has got it wrong and the other one has got it right." In section 8 we shall see how such a symmetry, in a sense, may be broken.

---

[18] For a clear explanation and representation of a simplified TT in a Gödel's universe see Smith and Oaklander (1995, 199-201).
[19] Note that there is no unique axis of rotation: in Gödel's universe *any* observer believes he/she lie at the centre of the universal rotation.

Lastly, it must be remarked that in Gödel's world the violation of chronology is *intrinsic*: chronology cannot be restored by "unwinding" the CTCs into open curves. In other words, the time structure of the Gödelian universe is not closed on a global scale as, contrariwise, is, for instance, the cyclical time structure of a universe consisting of Minkowski spacetime (topologically) rolled-up along the time axis into a cylindrical shape (as in fig. 4).

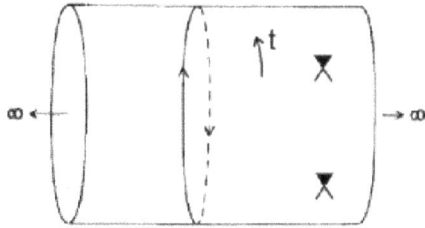

Fig. 4. Rolled-up Minkowski spacetime.

In this last case, in fact, all CTCs can be removed by cutting the hypercylinder, that is to say by breaking off their continuity (and then their closure) in a given line. And this possibility is theoretically given by the fact that Einstein field equations determine only the metric and *not* global topology of spacetime, therefore one is not forced by General Relativity to accept the possibility of CTCs simply because one is not forced to accept, in our last case, that global cylindrical topology. On the contrary, Gödel's spacetime *must* be interpreted in terms of a geometry *intrinsically* containing CTCs, since they do *not* derive from "artificial" topological identifications and consequently they cannot be removed by changing the global topology.[20]

---

[20] Even though a four-dimensional rolled-up Minkowski spacetime and a Gödel spacetime have different topological and metrical structures, they both have a CTC passing through *every* point, and for each point p it results that $I^+(p) = M$, namely the future (but also the past) light cone of any point encompasses the whole universe. For this reason in both universes it is possible to travel, by following the right sort of pathway, between any points. However, one significant difference between these two spacetimes is that a TT in Gödel's universe does not necessarily traverse the whole history of the universe, whereas a TT in the rolled-up spacetime does.

6. Advantages and disadvantages of Gödelian over Wellsian time travel

Earman (1995, par. 6.2) points out that GTT, at first sight, seems to have the following three important advantages over WTT:

1) GTT would seem to count as *physically possible*, in the sense that it is compatible with accepted physical laws, at least as regards General Relativity;
2) GTT is *not* open to a rereading on which no TT takes place;
3) GTT does *not* involve backward causation (at least not in the form arising in WTT).

The problem is that these three advantages are not so strong and "free" as they appear. As regards the first point, Earman himself underlines that the conception of physical possibility as compatibility with physical laws could be judged naïve.[21]

As regards advantages 2) and 3), they are more significant, being, as it were, gained at an expensive price, in the sense that one could object that GTT is not a TT after all:

> Gödelian time travel does not deliver time travel in the sense wanted since Gödelian time travel so-called implies that there is no time in the usual sense in which to 'go back.' In Gödel universe, for example, there is no serial time order for events, since every spacetime point p chronologically precedes itself; nor is there a single time slice which would permit one to speak of the Gödel universe at a given time. (Earman 1995, 164)

Thus the problem which in a sense weakens, or even risks to invalidate, the very status of "time travel," is that GTT does not occur in a global temporal background. This is an argument accepted by Earman, but in the meantime he affirms that GTT, even if it is not properly "time travel," is in any case a worthy object of investigation over and above the labels we stick on it. In this way, he seems to subscribe to such an "incompleteness" of the GTTs, thus assigning them a sort of weakened status of TT. This defect does not apply to WTTs, but according to Earman the latter have no space in physics.

---

[21] This is an issue we cannot pursue here. But we just want to note that even if, on the one hand, the mere compatibility of GTT with such a well-confirmed theory as General Relativity is not enough to assign definitely a "license" of physical possibility, on the other hand, it is by no means secondary that WTT significantly jars with physics (at least the current physics).

We disagree with Earman, in the sense that we think it is possible to recover a proper sense of TT also for a large class of GTTs, namely those occurring in particular bounded regions of spacetime. But before explaining in which way this idea may be supported, a deeper look is necessary at the status itself of TT as emerges in both the Wellsian and the Gödelian case.

7. *Wellsian and Gödelian status of time travel*

In our opinion, WTT (despite its numerous indigestible characteristics) may be seen, in a sense, as the "true" TT, insofar as it is a journey *through* time, i.e. a "movement" *backward* (or forward) in a time, which constitutes a global ordered temporal background. The simple reason for this state of affairs is, as already seen, that WTT is a journey in the context of a Newtonian or Minkowskian universe, namely in a 3D Euclidean spatial manifold that changes along an inexorable all-embracing 1D arrow of time. So here there are single time slices that allow us to speak of the universe at a given *objective* time. GTT, instead, is a journey set in the context of an "Einsteinian" universe, i.e. in a 4D curving spacetime continuum, in which time has the character of a spatial dimension, in the sense that there can be local variations or "warps" in its structure, and in which, in general, no global decomposition of spacetime in space and time is possible. Accordingly, GTT does not avail itself of that foliation of spacetime that represents an ubiquitously valid temporal frame and, in our opinion, necessary in order to speak of an actual journey *in* time. There is here only a personal time directed *always* towards the *local* future of the traveler. And, of course, it is not even possible to find, in a CTC, a sort of "backward" part, i.e. a portion of the curve along which one can sensibly say that the trip is taking place back *in* time, precisely because there is no external time frame.

It results that a GTT is not strictly a TT but is, as it were, a journey "of" (space)time insofar as a CTC instantiates that sort of spatiotemporal constraint which binds the advance of the traveler, and therefore also the return to the starting point or, in the case of an *almost* CTC, to a near point. Roughly speaking, we may say that, through a GTT, one goes *forward* to the past. But the point is that, when one thinks about TTs it seems much more natural to refer to a "form of displacement" by which one goes *back* to the past, namely to something exactly similar, at least from this point of view, to a WTT.[22]

In short, if we want to draw some general conclusions on the status of TT regarding these two types seen so far, it seems that, also in the light of the preceding sections, neither a WTT nor a GTT are particularly satisfactory in order to be a *proper* TT. The latter, indeed, should approach the physical laws in a consistent way, therefore preserving genidentity but not implying backward causation (as GTT does and WTT does not), and, in the meantime, it should also possess that Wellsian "return *back* to the past" status typical of WTT but absent in the GTT. Now, while WTT fails, for the reasons seen above, to overcome the drawbacks by getting rid of backward causation and its related problems, GTT can instead be "improved," i.e. it can recover, at least partially, that missing status previously underlined. In order to provide it with this last feature, we need to blend two ingredients: some particular spatiotemporal topologies allowing CTCs, and an ordered global time. These are the matters, respectively, covered in the next two sections.

8. Chronology violating regions

Gödel's universe was the first forerunner of temporally anomalous spacetimes.[23] Actually, General Relativity allows other solutions exhibiting CTCs and then other opportunities for TTs. The first one we want to outline briefly is the Taub-NUT spacetime. Fig. 5 (from Earman et al. 2009) shows a simpler bidimensional version (with topology $S \times \mathbb{R}$) given by Misner. In this spacetime, as happens in Gödel's, the CTCs are due to the "tilting" of light cones.

---

[22] The semantic difference hidden in these two locutions is simple: in the Wellsian case the term "back" (or also "forward") is thought to be referred to a global external time; instead, in the Gödelian case, the term "forward" is obviously referred to the traveler's proper time.

[23] Actually, in 1937, van Stockum found an exact solution (consisting of an infinite rapidly rotating cylinder of dust, surrounded by a vacuum) to the Einstein field equations containing (at the exterior of the cylinder) CTCs. But the existence of this latter specificity was realized only many years later.

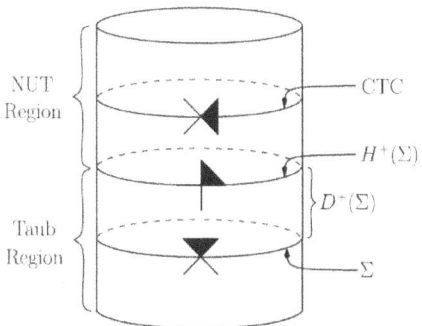

Fig. 5. Misner's 2-dimensional version of Taub-NUT spacetime.

However, Taub-NUT spacetime, unlike Gödel's, shows that rotating matter is not a necessary condition for the formation of CTCs in the general relativistic setting, since Taub-NUT spacetime is a *vacuum* solution to the Einstein field equations. But it also shows another, and more important for our purposes, fact: CTCs do not concern the *whole* spacetime, whereas in those of the rolled-up Minkowski and Gödel they are *everywhere*. Indeed, one finds that the Taub region (the lower part of the cylinder)—interpreted as a cosmological solution with homogeneous but nonisotropic space sections (the t = constant hypersurfaces are space-like)[24]—evolves into the NUT region (the upper part), the only one to contain CTCs.[25]

In other general relativistic spacetimes, this separation between regions that have CTCs and others that do not is still more evident, in the sense that the latter—the *normal regions*, to use Arntzenius and Maudlin's (2002) term—are almost the entire spacetime whereas the former—call them *TT regions* or *chronology* (or *causality*) *violating regions*—are more localized, that is to say they occupy just small, or in any case limited, spatiotemporal portions.

---

[24] See Misner 1967.

[25] In the future of the partial Cauchy surface Σ the light cones begin to "tip over" so that the surfaces of homogeneity, in the future domain of dependence $D^+(\Sigma)$, become more and more lightlike until, eventually, in the future Cauchy horizon $H^+(\Sigma)$, the generation of a closed null curve indicates that CTCs are on the verge of forming; and this actually happens in the NUT region. See Hawking and Ellis (1973, 170-178), and Earman et al. 2009.

For instance, the *Deutsch-Politzer spacetime*, represented in fig. 6 (from Wüthrich 2009), is an exact solution to the Einstein field equations with a local "handle" (it is a combination of a curved cylinder and a plane with 2 holes in the Minkowski spacetime) containing CTCs. This spacetime loses its "good" causal behavior in the course of its evolution: it is identical with empty Minkowski spacetime, thus it is normal everywhere, until, at a certain point, it develops the handle region in which all time loops are confined. The consequence of such a structure is that a traveler that passes into the handle will enter a chronology violating region in a way compatible with the current state of the universe.

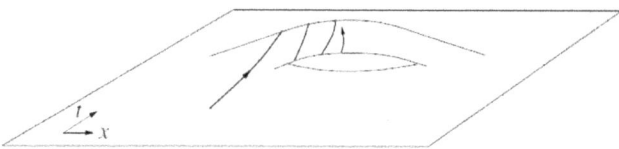

Fig. 6. The Deutsch-Politzer spacetime.

On the other hand, Deutsch-Politzer spacetime is exactly as "physical" as the Minkowski spacetime, being a solution with the same (zero) source and initial conditions. However, this spacetime gives no clue as to what might force the Minkowski spacetime to evolve into this structure containing CTCs instead of just remaining Minkowskian. In other words, it is not clear how the state on the portion of the Deutsch-Politzer spacetime that does not contain CTCs, insofar as it is identical to the Minkowski spacetime, could be responsible for the appearance of the chronology violating region. The result is that CTCs are possible, but not guaranteed. Accordingly, Earman and Wüthrich (2004) explicitly state that Deutsch-Politzer spacetime is not a plausible candidate for a time machine spacetime.

Even if the solutions to Einstein field equations seen so far probably do not represent the actual universe, their simple mathematical "unfolding" raises important foundational questions: Is it (at least in principle) possible to create CTCs, by local rearrangements of the distribution and flow of matter (all compatible with General Relativity), giving in such a way a chance to local TTs? Are there natural processes which lead to the formation of CTCs in *confined* regions of our universe? The first question concerns the possibility to manufacture (by future technology, who knows!) what is called, in a modern sense, a *time machine*.[26] The second one regards the existence of *natural* phenomena enabling TTs through the "spontaneous generation" of CTCs. It is the last question in particular that interests us here. Let us remember once more that speaking of confined regions means referring to bounded, though possibly large, local regions within a cosmos whose structure is obviously on the whole "normal" (i.e., devoid of CTCs). There is no shortage of examples that seem to allow a positive answer to the second question, even if the debate on their physical possibility is still truly alive. We shall briefly consider only one example: the Kerr-Newman spacetime.

The *Kerr-Newman metric* describes the geometry of the spacetime around a rotating, non-spherical, charged collapsed object (in short, a particular kind of black hole). This is the only metric, of the four different types describing black holes, that contains a *naked singularity*, i.e. a singularity not hidden by a boundary (event horizon). Such a singularity is thus theoretically observable from the outside of the black hole, and matter and radiation can both fall in and come out through it (the black hole is not so black!). Actually, only one type of the Kerr-Newman metric allows a naked singularity. Indeed, this metric is characterized by two parameters: the angular momentum a per unit mass, and the mass m of the collapsed object. Only the case $a^2 > m^2$, in which the black hole spins sufficiently fast and the events horizons vanish, produces a naked singularity in the form of a *ring singularity*, namely a circular line (placed at $r = 0$, with mass, electromagnetically charged, rotating around its axis of symmetry) through which matter may pass without destruction. The point is that the spacetime nearby the ring singularity is a chronology violating region, because the rotating charged ring-shaped singularity produces CTCs.[27] In this way, the lack of an event horizon implies that the CTCs of such a region are exposed to the rest of the universe, thus allowing a time traveler to presumably embark upon his/her journey. Just to get an idea, fig. 7 (from Thorne 1993, 298) shows the hypothetical chronological structure of a spacetime

---

[26] A time machine is, to put it crudely, a device that produces CCCs in a spatiotemporal region (to the future of its operation) where otherwise none would have existed.

[27] The physical reason, roughly speaking, is that the rotation of matter (i.e. the rotation of the gravitational source) drags spacetime around itself (called the *frame-dragging effect*). Andréka et al. 2008 maintain that a supposed time traveler should orbit in the direction opposite to the rotation of the black hole insofar as the proper time of the CTCs counter-rotates, contrary to popular belief, against the rotational sense of the ring-singularity.

that might result from an axially symmetric contraction of a finite sized, rotating body.

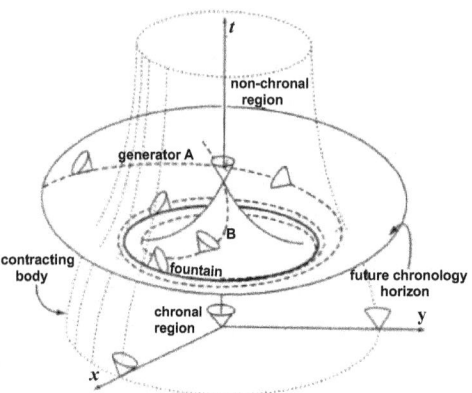

Fig. 7. The chronological structure in a rapidly rotating and contracting body's equatorial plane.

Roughly, this is what should happen: at early times, when the body is large, there are no CTCs (such a spacetime is termed *chronal*); at late times, when the body has settled down into its smaller state, light cones tilt; the chronal region ends and in the future chronology horizon two kinds of CTCs are created: generators A, i.e. null geodesics spiralling outward and remaining always on the horizon, and generators B, instead spiralling inward, and then leaving the horizon at the tip of the "Mexican hat" of fig. 7. The moral is: as there are no event horizons, one can travel from our universe to the past falling through the nonchronal region, spiralling around the axis of symmetry for a long enough time and finally coming back to chronal region again at the same time (or at an earlier one) as the departure.

Of course, all this matter is very speculative and sounds like daring science fiction. Although the existence of black holes in our universe is generally accepted, there remain doubts on the real existence of such incredible objects as the fast rotating Kerr-Newman black holes.[28] And any-

---

[28] The slow-Kerr-Newman black hole (i.e., the case in which $a^2<m^2$), or the Kerr black hole (i.e., the uncharged case), may perhaps be judged more physical. In such cases,

way there would be several challenges to the practicability of these TTs (inaccessibility of these regions of the universe, enormous energy requirements, etc.). Accordingly, not all researchers are ready to judge these natural time machines of cosmic size as part of our reality, and some of them are still attempting to extract from General Relativity theorems (the so-called *no-go theorems*) establishing the impossibility to operate whatever time machine.[29]

However, as Earman et al. (2003) have argued, these theorems have not definitely ascertained such an impossibility. In fact, other eminent physicists (e.g., Morris and Thorne 1988) have put forward the possibility that an advanced civilization might produce *artificial* machines for backward TTs (by means of traversable spacetime wormholes). Furthermore, behaviors similar to the solutions so far seen occur in other geometries,[30] thus, in general, CTCs, though still very far from participating in our daily lives, cannot be easily discarded as unphysical scenarios. Moreover, in these pages our main concern is not the practicability of TTs. However if we take General Relativity theory seriously, we have to weigh the plausibility either that in our universe, in one way or another, chronology violating regions already exist, or that the progressive development of our universe could generate them in a natural (or artificial: with the participation of some advanced civilization) way.

That being said, let us come back to Kerr-Newman black holes, in order to reason upon another point that perhaps could be extended also to other cases of local TTs. The fact that this spacetime is asymptotically flat[31] pushes us towards other considerations concerning the status of these

---

there is still a chronology violating region, but this time CTCs, the ring singularity being "clothed," are surrounded by the event horizon, so they are not exposed to our universe. Nothing falling through this horizon can ever get back out, thus they cannot be used for TT: a one-way trip towards the ring clothed singularity is allowed, but sadly no return!

[29] The most famous and discussed critical statement in this regard belongs to Stephen Hawking. It is the so-called *chronology protection conjecture*: "It seems that there is a Chronology Protection Agency which prevents the appearance of closed timelike curves and so makes the universe safe for historians" (Hawking 1992, 603). Simply put, it states: nature abhors time machines!

[30] For a minimum list of these temporally ill-behaved spacetimes (and relative references), see Visser (2003, 162).

[31] I.e. light cones start to tilt as they begin moving towards the black hole, while, at large distances, they are not influenced by its gravitational attraction, so that we can safely assume that they are "straight" like in Minkowski spacetime.

local TTs. Usually, what happens in a chronology violating region is presented saying that: "Light cones may be so much tilted with respect to a local coordinate time direction, that there exist timelike and null trajectories which run backwards with respect to the time coordinate causing the local causal future to overlap with what would have been the causal past in a flat spacetime" (de Felice 1995, 101). In a word, the local trip into the past is figured out by referring it to a hypothetical *local* "flat" time. The question then is: why not extend, at least conceptually and with the necessary precautions, this temporal frame to a *global* coordinate time direction? On the other hand, a sort of similar extension seems to be adopted implicitly by de Felice and collaborators themselves in their in-depth researches on the causal structure of the Kerr and the Kerr-Newman spacetimes. In fact, in order to have a travel into the past for a timelike path entering the chronology violating region, they explicitly take as reference a stationary observer at infinity (by which they mean sufficiently far away from the metric source), so that the time coordinate of this path runs backwards just *with respect to* infinity.[32] In particular, they distinguish two conditions for the phenomenon of the violation of causality to take place (again, with respect to a stationary observer at infinity) in a Kerr-Newman spacetime. The first, a necessary condition, is the simple entering the chronology violating region. The second, a more stringent (and sufficient) condition that has to be satisfied on the path entering that region, is explained in this way:

> There must exist a timelike curve which connects two points at (positive) infinity such that the gradient of the coordinate time (so chosen as to represent the proper time of an observer at infinity) relative to an affine parameter on the curve vanishes on a curve point; this point will be called the chronology violating point of the curve and the curve itself a chronology violating curve.[33] (de Felice 1981, 225)

This means that the traveler on the path inside the chronology violating region meets a turning point (where the time component of the four-velocity $\dot{t}$ vanishes and then changes sign),[34] after which the traveler starts to return to the asymptotic regions at the same or earlier time as that of departure. In other words, at that turning point the traveler enters, so to speak, the

---

[32] See, for instance, de Felice et al. (1980, 3635); de Felice and Calvani (1979, 336).
[33] This criterion is more explicitly stated in Calvani et al. 1978.
[34] In fact, they call a causality violating region a region where the condition $\dot{t} \leq 0$ holds.

"backward" part of the CTC and hence starts moving into the past, or rather, towards what a distant observer considers in this way in relation to the flow ahead of the traveler's proper time. Note that here the time coordinate *t*, as de Felice and Calvani point out, "has an intrinsic physical meaning only where the metric is nearly flat, being the proper time of the static observers there;" and after they immediately add: "It is to these observers that the causality violation examined here has to be referred" (de Felice and Calvani 1979, 336), namely, it is these distant observers that give, so to speak, sense to the violation of the causality in the form of that local TT. However, it is important to repeat that from the viewpoint of the local Gödelian traveler there is no particular point at which the "backward phase" of the journey begins, since, on the traveler's CTC, the journey always moves *forward* in terms of proper time.

Now, referring the causality violation to these distant stationary observers is nothing more than referring the becoming[35] of that local GTT to a reference system that, at another scale, namely at a cosmological level, constitutes the so-called *substratum* whose main characteristic is to be able to define a cosmic time. We shall analyze this in the next section, but, before concluding the present one, still more clarification can be gleaned from the following words of de Felice in which he sums up the salient moments of a TT inside a black hole chronology violating region:

> Let a time coordinate be chosen so to coincide with the proper time of an observer at infinity. If a test particle, initially at infinity [...], starts a journey towards the singularity, then, in order to go back arbitrarily close to the starting point (nearly the same place) but *at the same or earlier time* of departure, it *has* to enter the kernel [i.e., the small domain nearby the singularity] and travel there for a sufficiently large value of its proper time on a trajectory which runs backwards with respect to the coordinate time. Thus, the observer who first launched the particle, will directly experience the paradoxical interaction of the particle at the end of the journey with its younger self before it left.[36] (de Felice 1995, 101; his italics)

---

[35] Here "becoming" is in a broad sense, however it might also be used, according to Savitt 2005, in the more strict senses of "temporal passage" or "flow of time" proper of the philosophy of time.

[36] Interestingly, de Felice 2008 illustrates what might be the observational implications, somewhere in our universe, of a cosmic time machine. It is conjectured that such a machine might be the source of fast varying and highly energetic events like *gamma ray bursts*.

9. Weyl's principle and the "Wellsianization" of GTT

At the basis of the standard model of cosmology there is an assumption firstly proposed in 1923, and better formalized in 1930, by Hermann Weyl, called in fact *Weyl's principle* (or *postulate*). In a modern formulation it states: "The worldlines of galaxies form a 3-bundle of non-intersecting [and diverging] geodesics orthogonal to a series of spacelike hypersurfaces" (Narlikar 2010, 230).

It makes more sense to refer this frame to the center of mass of clusters of galaxies rather than to single galaxies because it is this center that seems to follow the Hubble expansion pattern quite closely. Although clusters form a discrete set, we can extend it to a continuum by a smooth-fluid approximation, a reasonable idealization valid only at large scales where clusters may be regarded as the particles, the so-called *fundamental particles*, out of which this fluid is made. Fundamental particles, when regarded as mere geometric points, constitute the kinematic *substratum*. At each point, crossed by only one geodesic, a locally inertial reference frame, along with its own *fundamental observer* equipped with a clock, can be defined.

An important assumption of Weyl's principle, not evident from the aforementioned Narlikar's formulation, is that this cosmic geodesic system is thought to have been *causally interconnected* since its origin.[37] Weyl's principle, indeed, stipulates that the large scale *mean* motion of fundamental particles is highly streamlined (no randomness in their motions, no collisions among particles, no vorticities in their trajectories) *except* at a singular point (the "origin") in the past (and possibly, but not necessarily, at a similar singular point in the future). This common origin and this regularity of the geodesics allow the geodesics themselves and their becoming to provide *natural synchrony calibration* for all events (the intersection defines the zero of time).[38] This guarantees that spacetime can be foliated in a sequence of time slices orthogonal to the bundle, namely spacetime is decomposed in space plus time, an all-embracing coordinate cosmic time t that is the relativistic analogue of the Newtonian time.

---

[37] Weyl's principle is sometimes also called *Weyl's causal postulate* or *Principle of common origin*.

[38] For example, clocks could be synchronized by the exchange of light signals between fundamental observers who could agree to set their clocks to a standard time when, e.g., the universal homogeneous density reaches some given value.

In our isotropic expanding universe, therefore, it is possible to imagine a privileged class of fundamental observers (those associated with the mean smeared-out motion of clusters) constituting an extended inertial system, thus these observers are freely falling along their worldlines, i.e. in expansion with the cosmic fluid, and are observing the universe as spatially isotropic and homogeneous, and are measuring as their proper time the coordinate cosmic time that marks the becoming of the universe as a whole.[39]

This fact brings us back to what de Felice and collaborators said at the end of the last section. It now seems natural to refer local GTT to this class of observers constituting the substratum. In so doing we establish a sort of correspondence between local and global, i.e. between the traveling on the CTC inside the chronology violating region and the cosmic time measured by those external distant fundamental observers. Thus the traveling in the past is related, or rather is conceived thanks, to that global temporal background.

It is now evident, we presume, in which sense a local GTT assumes a "Wellsian aspect." Indeed, even if in a Gödel's universe the time order of a CTC cannot be compared with the temporal orderings of its neighboring or distant observers' worldlines, insofar as there is no global unique direction of time, in a universe respecting Weyl's principle instead this can be done, at least in principle, for the time order of a local CTC. Such a curve, in fact, inhabits a world where a global temporal order with respect to the comparisons "later than" or "earlier than" exists. And this enables us to assign, also to travels on those local CTCs, a sort of objective temporal reference frame thanks to which the locution "travel *in* time" may acquire a valid meaning. This is exactly what happened for WTTs in Newtonian and Minkowskian spacetimes. For this reason a situation like the one portrayed by Richmond at the end of section 4, here no longer holds. In that Gödelian case the situation was perfectly symmetric but now, if observers A and B

---

[39] Spatial homogeneity and isotropy constitute the other fundamental principle of standard cosmology: the *cosmological principle*. Usually, it is this principle that is placed at the foundations of the *Friedmann-Lemaître-Robertson-Walker spacetime*, i.e. the metric considered describing the large-scale structure of our universe, whereas Weyl's principle is often more or less tacitly assumed or included in the mathematical formulation of the cosmological principle itself. But we consider Weyl's principle more fundamental than (a precondition for) the cosmological principle (see Rugh and Zinkernagel 2011, and also Macchia 2011, chapt. 3), therefore we introduce the former from the outset, and, successively, we apply the latter to the hypersurfaces of simultaneity (for a mathematical "translation" of this approach see Narlikar 2010, 229-232).

are taken to be, respectively, a fundamental observer and a traveler in a local CTC, one obtains that it is A that—to use Richmond's expression—"has got it right." When Weyl's principle holds, indeed, one has a *prevailing* temporal order, the global one, so that A may refer the time measurements to a sort of objective temporal setting.[40]

As regards what happens inside the chronology violating region, obviously it is not possible to extend the cosmic time into its interior because its topological structure prevents us from doing it. But our idea is simply that a local CTC-traveler *is* a traveler *in* time because, since the parameter labeled "time" in his/her CTC (his/her proper time) may be associated with the external "genuine" cosmic time, the traveler actually goes *back* to the past relatively to an objective temporal background.[41]

A possible criticism to our reasoning might be based on the following considerations. The vorticity-free claim constitutive of Weyl's principle is equivalent to the requirement of stable causality for the spacetime because, as we have told in section 3, the possibility to define a global cosmic time is given by the fact that spacetime is causally stable or, still better, globally hyperbolic. In Malament's words: "In a cosmological model, non-rotation of the major mass points is precisely the necessary and sufficient condition for there to exist a natural notion of simultaneity relative to their worldlines" (Malament 1995, 263). He explains this crucial correlation between rotation and temporal structure with an intuitive illuminating metaphor: "Think about an ordinary rope. In its natural twisted state, the rope cannot be sliced in such a way that the slice is orthogonal to all fibers. But if the rope is first untwisted, such a slicing is possible. Thus orthogonal sliceability is equivalent to fiber untwistedness" (*ibid.*). For this reason a universe respecting Weyl's principle and a Gödel's universe are antithetic. Moreover, remember, again thanks to section 3, that causal stability reflects also a complete absence of CCCs, more precisely that no perturbations in the metric can produce them, so Weyl's principle automatically, *by definition*, excludes the possibility to have CTCs *anywhere*. Hence, how might one reconcile a causal stability that implies cosmic time

---

[40] Clearly, if the traveler moves along a CTC at the velocity of light, the worldline would be perpendicular to cosmic time, so that the journey could not be considered Wellsian in any sense, since from that point of view it would be instantaneous. In spite of this if the traveler moves more slowly the Wellsian character re-emerges.

[41] We do not pretend that such an association be an isomorphism, but we claim that every possible association respects the direction of the cosmic time order, even if it is not possible to *uniquely* assign the cosmic time order to the traveler in the chronology violating region (we thank Chris Smeenk for this last observation).

through the absence of CCCs with the supposed possibility of local GTTs in the chronology violating regions? In other words, in terms of Cauchy initial value problem, how can one manage to combine the fact that if there are CTCs no Cauchy surfaces can exist (and then Weyl's principle cannot hold given that it splits spacetime into Cauchy hypersurfaces)?

Mathematically, even if in specific cases the technical details could be really complicated, such a mutual exclusiveness can be side-stepped on the basis of the following consideration: CTCs are locally important, but if one considers a sufficiently coarse-grained model of the universe (as cosmology necessarily does), they become irrelevant. That is to say: even if we cannot have a Cauchy surface (all over the spacetime) and then the certainty that in the future geodesics do not intersect, in reason of the solution to the traditional initial value problem, we can recover a *partial* Cauchy surface that entirely inhabits the normal regions of spacetime and that admits the development of CTCs in the future, though allowing the definition of a partial cosmic time.[42]

From a physical point of view, things appear simpler: even if we have two metrics that seem to clash—the Friedmann-Lemaître-Robertson-Walker metric, valid at very large scales for a homogeneous and isotropic universe, and the Kerr-Newman metric, valid at small scales in the surroundings of a particular black hole—actually, they both can be framed in the same picture precisely because they are not incompatible. Indeed, the first metric does not pretend to describe what physically happens to the universe at small scales (which is anything but homogeneous and isotropic), and the second one does not claim to embrace the universe as a whole.[43] In short, we have two descriptions of the universe—a coarse-grained one (involving the global cosmic time) and a fine-grained one (involving local traveler proper time)—which are compatible exactly because they reflect two spatial scales naturally different but in physical compatible coexis-

---

[42] A *partial Cauchy surface* is a time slice intersected by any causal curve *at most* once (remember: a Cauchy surface is instead intersected *precisely* once). "Partial" means that only a portion of the future history of the spacetime can be predicted from this surface, that is one has to exclude that part of spacetime in which the chronology violating region will form.

[43] A comparison of the dimensions at stake is completely hazardous. But, just to have an idea, think that a single fundamental particle abstractly represents (the center of mass of) an entire cluster whose typical diameter may be roughly considered as 10^{23} m, whereas the size of the supposedly biggest black hole (a supermassive one) is thought to arrive at most at 10^{10} m. So the difference in orders of magnitudes is no less than 13!

tence. And in fact cosmology tells us that black holes inhabit our universe (supposedly homogeneous and isotropic).

10. Conclusion

Whether chronology violating regions are physical possibilities for our universe is still an open question. However, there are solutions to the Einstein field equations which generate CTCs with certain ease. This implies that if General Relativity is to be taken seriously (but a theory of quantum gravity that we are still waiting for could modify the situation), then also the possibility of TT, in the form of local CTC, must be taken seriously. In a universe like ours, furthermore, in which Weyl's principle holds, it is possible to define a cosmic time, namely it is possible to assume at very large scales the global validity of a sort of Newtonian temporal order. It is to this order that the local temporal order of the GTT occurring in a chronology violating region has to be referred. This enables us to conclude that the Gödelian traveler's proper time, from a fundamental observer's point of view, flows backwards with respect to cosmic time. In this way, one recovers that Wellsian, and more natural and objective, sense of travel *through* time also for the mostly physical journey on CTCs.

REFERENCES

Andréka, H., Németi, I., and Wüthrich, C. (2008). A Twist in the Geometry of Rotating Black Holes: Seeking the Cause of Acausality. *General Relativity and Gravitation*, 40(9), 1809-1823.

Arntzenius, F., and Maudlin, T. (2002). Time Travel and Modern Physics. *Time, Reality and Experience*, ed. C. Callender, Cambridge: Cambridge University Press, 169-200.

Calvani, M., de Felice, F., Muchotrzeb, B., and Salmistraro, F. (1978). Time Machine and Geodesic Motion in Kerr Metric. *General Relativity and Gravitation*, 9(2), 155-163.

Carnap, R. (1967). *The Logical Structure of the World and Pseudoproblems in Philosophy*. Berkeley: University of California Press.

de Felice, F. (1981). Timelike Nongeodesic Trajectories which Violate Causality. A Rigorous Derivation. *Il Nuovo Cimento*, 65B(1), 224-232.

de Felice, F. (1995). Cosmic Time Machines. *Lecture Notes in Physics*, 455, 99-102.

de Felice, F. (2008). Naked Singularities, Cosmic Time Machines and Impulsive Events. arXiv:0710.0983v1.

de Felice, F., and Calvani, M. (1979). Causality Violation in the Kerr Metric. *General Relativity and Gravitation*, 10(4), 335-342.

de Felice, F., Nobili, L., and Calvani, M. (1980). Charged Singularities: the Causality Violation. *Journal of Physics A: Mathematical and General*, 13, 3635-3641.

Earman, J. (1995a). Recent Work on Time Travel. *Time's Arrows Today*, ed. S. F. Savitt, Cambridge: Cambridge University Press, 268-310.

Earman, J. (1995b). *Bangs, Crunches, Whimpers, and Shrieks: Singularities and Acausalities in Relativistic Spacetimes*. New York: Oxford University Press.

Earman, J., and Wüthrich, C. (2004). Time Machines. *Stanford Encyclopedia of Philosophy*, ed. E. N. Zalta, URL: <http://plato.stanford.edu/entries/time-machine>.

Earman, J., Smeenk, C., and Wüthrich, C. (2003). Take a Ride on a Time Machine. *Reverberations of the Shaky Game: Festschrift for Arthur Fine*, eds. R. Jones and P. Ehrlich, Oxford: Oxford University Press.

Earman, J., Smeenk, C., and Wüthrich, C. (2009). Do the Laws of Physics Forbid the Operation of Time Machine? *Synthese*, 169, 91-124.

Fano, V., and Tassani, I. (2002). *L'orologio di Einstein. La riflessione filosofica sul tempo della fisica*. Bologna: CLUEB.

Gödel, K. (1949). An Example of a New Type of Cosmological Solutions of Eintein's Field Equations of Gravitation. *Reviews of Modern Physics*, 21, 447-450.

Hawking, S. W. (1968). The Existence of Cosmic Time Functions. *Proceedings of the Royal Society of London. Series A, Mathematical and Physical Sciences*, 308(1494), 433-435.

Hawking, S. W. (1992). The Chronology Protection Conjecture. *Physical Review*, D46, 603-611.

Hawking, S. W., and Ellis, G. F. R. (1973). *The Large Scale Structure of Space-Time*. Cambridge: Cambridge University Press.

Heller, M. (1990). Time and Causality in General Relativity. *The Astronomy Quarterly*, 7, 65-86.

Horwich, P. (1987). *Asymmetries in Time*. Cambridge: The MIT Press.

Lewis, D. (1986). The Paradoxes of Time Travel. *American Philosophical Quarterly*, 13, 145-152.

Macchia, G. (2011). *Fondamenti della cosmologia e ontologia dello spaziotempo*. PhD Dissertation in Humanistic Sciences, University of Urbino (Italy).

Malament, D. (1984). 'Time Travel' in the Gödel Universe. *Proceedings of the Biennial Meeting of the Philosophy of Science Association*, Vol. 2: Symposia and Invited Papers, 91-100.

Malament, D. (1995). Introductory Note to *1949b. K. Gödel. Collected Works*, eds. S. Feferman et al., Vol. 3, Oxford: Oxford University Press, 261-269.

Meiland, J. (1974). A Two-Dimensional Passage Model of Time for Time Travel. *Philosophical Studies*, 26, 153-173.

Misner, C. W. (1967). Taub-NUT Space as a Counterexample to Almost Anything. *Relativity Theory and Astrophysics: I. Relativity and Cosmology, Lectures in Applied Mathematics*, Vol. VIII, ed. J. Ehlers, Providence: American Mathematical Society, 160-169.

Morris, M. S., and Thorne, K. S. (1988). Wormholes in Spacetime and Their Use for Interstellar Travel: A Tool for Teaching General Relativity. *American Journal of Physics*, 56, 395-412.

Nahin, P. J. (1999). *Time Machines*. 2nd ed. New York: AIP Press.

Narlikar, J. (2010). *An Introduction to Relativity*. Cambridge: Cambridge University Press.

Putnam, H. (1962). It Ain't Necessarily So. *Journal of Philosophy*, 59(22), 658-671.

Reichenbach, H. (1957). *The Philosophy of Space and Time*. New York: Dover Publications.

Richmond, A. (2006). Time Travel and its Philosophical Problems. URL: <http://vidiowiki.com/media/paper/wmax75e%20wmax75e.pdf>.

Rugh, S., and Zinkernagel, H. (2011). Weyl's Principle, Cosmic Time and Quantum Fundamentalism, eds. D. Dieks et al., *Explanation, Prediction, and Confirmation*, Dordrecht: Springer, 411-424.

Savitt, S. (2005). Time Travel and Becoming. *The Monist*, 88(3), 413-422.

Smart, J. J. C. (1955). Spatialising Time. *Mind*, New Series, 64(254), 239-241.

Smeenk, C., and Wüthrich, C. (2011). Time Travel and Time Machines. *The Oxford Handbook of Philosophy of Time*, ed. C. Callender, Oxford: Oxford University Press, 577-630.

Smith, Q., and Oaklander, L. N. (1995). *Time, Change and Freedom*. London: Routledge.

Thorne, K. S. (1993). Closed Timelike Curves. *General Relativity and Gravitation, 1992. Proceedings of the 13th International Conference on General Relativity and Gravitation*, eds. R. J. Gleiser, C. N. Kozameh and O. M. Moreschi, Bristol: Institute of Physics Publishing, 295-315.

Visser, M. (2003). The Quantum Physics of Chronology Protection. *The Future of Theoretical Physics and Cosmology: Celebrating Stephen Hawking's 60th Birthday*, eds. G. W. Gibbons, E. P. S. Shellard and S. J. Rankin, Cambridge: Cambridge University Press, 161-176.

Weingard, R. (1979b). General Relativity and the Conceivability of Time Travel. *Philosophy of Science*, 46, 328-332.

Weyl, H. (1923). *Raum-Zeit-Materie*. 5th ed. Berlin: Julius Springer.

Weyl, H. (1930). Redshift and Relativistic Cosmology. *Philosophical Magazine*, 9, 936-943.

Wüthrich, C. (2003). Does Modern Physics Permit the Operation of Time Machine? URL.: <http://aardvark.ucsd.edu/grad_conference/wuthrich.pdf>.

Wüthrich, C. (2009). Time Travel. URL: <http://philosophy.ucsd.edu/faculty/wuthrich>.

About the authors

Miloš Adžić is teaching assistant at the Department of Philosophy, Faculty of Philosophy of the University of Belgrade. His main research interests include logic, both philosophical and mathematical.

Miloš Arsenijević. Professor of philosophy at the University of Belgrade. Fellow of the Center for the Philosophy of Science, University of Pittsburgh. Former lecturer of philosophy, University of Heidelberg, and mathematics, University of Maryland. Visiting Professor at the University of Graz. Visiting university talks at Berkeley, Los Angeles, Santa Barbara, Pittsburgh, Montreal, Oxford, London, York, Leeds, Paris, Bielefeld, Karlsruhe, Darmstadt, Dortmund, Oslo and others. Published two books on space and time in Serbian, and dozens of articles (in *Analysis*, *Erkenntnis*, *Apeiron*, *Journal of Applied Logic*, *Grazer philosophische Studien*, etc.) and essays included in the collections of *MIT Press*, *Kluwer*, *Klostermann*, *Springer*, *Ontos*, etc. Main research interests: metaphysics, logic, philosophy of mathematics and physics.

Dennis Dieks is professor of the philosophy and foundations of the natural sciences at Utrecht University, The Netherlands. He is a member of the Royal Netherlands Academy of Sciences and member of the boards of several international professional organizations, both in physics and philosophy. He is also editor of the journal *Studies in the History and Philosophy of Modern Physics* (Elsevier) and associate editor of *Foundations of Physics* (Springer). He has published widely on subjects relating to space and time, the foundations and philosophy of quantum theory, probability and the philosophy of physics in general.

Vincenzo Fano is associate professor of logic and philosophy of science at the University of Urbino "Carlo Bo". His main research interests are in philosophy of physics and epistemology. He published on *Brentano Studien*, *Meinong Studies*, *Axiomathes*, *Studia Leibnitiana*, *Foundations of Physics*, *Disputatio* and *Epistemologia*. His most recent book is *I paradossi di Zenone*, Carocci, 2012.

Pierluigi Graziani is currently a Post-Doc in Logic at the University of Urbino "Carlo Bo", Italy. He works extensively in the following research areas: Structural Proof Theory and Structural Proof Analysis of Mathematical Theories; formal tools for the history of science; Mathematical School of Urbino. Graziani has published contributes on these topics in international journals, proceedings, dictionary, and co-edited books. He has organized international conferences on these topics and has also founded the "Lectiones Commandinianae" (Lectures on History and Philosophy of Mathematics). More details at: http://sites.google.com/site/grazianipierluigi/

Andrea Iacona is professor of logic at the University of Turin, Italy. His research interests are mainly in topics of philosophical logic, such as propositions, vagueness, future contingents, validity, counterfactuals and logical form. He has published papers on these topics in several journals, including *Erkenntnis*, *Dialectica*, *Synthese*, *Australasian Journal of Philosophy*, *American Philosophical Quarterly*. His books are *Propositions* (Name, 2002), *L'argomentazione* (Einaudi, 2005) and *Teoria della logica del prim'ordine* (Carocci, 2010).

Gregory Landini is professor of Philosophy at the University of Iowa. He is author of four books: *Frege's Notations; what they are and how they mean* (Palgrave/MacMillan, 2012), *Russell* (Routledge, 2011), *Wittgenstein's Apprentice with Russell* (Cambridge, 2007) and *Russell's Hidden Substitutional Theory* (Oxford, 1998). He has published many articles in the philosophy of logic and metaphysics. His teaching and research interests include modal logic, the foundations of mathematics, philosophy of mind, philosophy of language, and the history of analytic philosophy.

Giovanni Macchia is post-doc at the Department of Philosophy and at the Rotman Institute of Philosophy of the University of Western Ontario, Canada. His primary research interests concern the foundations of physics, especially the philosophical implications of General Relativity and the history and philosophy of modern cosmology. Other interests include time travels, metaphysics of time, philosophy of science. He is writing an introductory book for an Italian publishing company (Carocci editore) on the philosophy of cosmology, and a historical book for the Max Planck Institute of Berlin in which he analyses a debate from the London *Times* in 1932 on the expansion of the universe.

Domenico Mancuso teaches Mathematics in high school in Turin. He holds undergraduate degrees in Mathematics and Philosophy from the

University of Bologna, and a PhD in Philosophy (2007) from the universities of Urbino and Paris-Sorbonne. His main research topics are the free will problem, which has been the subject of his doctoral work, and a priori issues in the philosophy of time, especially time travel and McTaggart's paradox. He has published a paper on freedom and possibility in a recent issue of *Philosophical Forum*.

Edwin Mares is a professor of philosophy at Victoria University of Wellington, in New Zealand. He writes mostly on non-classical logic and has published in journals such as *Synthese*, the *Journal of Philosophical Logic*, and *The Journal of Symbolic Logic*. He is on the editorial boards of the *Australasian Journal of Philosophy*, *Studia Logica*, *The Review of Symbolic Logic*, and *Thought*. He has written three books: *Relevant Logic: A Philosophical Interpretation* (Cambridge University Press, 2004), *A Priori* (Acumen, 2011), and (with Stuart Brock) *Realism and Antirealism* (Acumen, 2007).

Francesco Orilia is professor of philosophy of language at the University of Macerata, Italy. His main research interests are in the theory of reference, logical paradoxes and ontology with a recent emphasis in the philosophy of time. He has published many papers in these areas in journals such as *dialectica*, *Journal of Philosophical Logic*, *Journal of Symbolic Logic*, *Notre Dame Journal of Formal Logic*, *Philosophical Studies*, *Synthese*. His most recent books are *Singular Reference. A Descriptivist Perspective* (Springer, Dordrecht, 2010) and *La filosofia del tempo* (Carocci, Roma, 2012).

www.ingramcontent.com/pod-product-compliance
Lightning Source LLC
Chambersburg PA
CBHW050856160426
43194CB00011B/2178